Advanced Techniques in Soil Stabilization for Road Construction

Advanced Techniques in Soil Stabilization for Road Construction

Jasmina Novakovic

ARCLER
PRESS

www.arclerpress.com

Advanced Techniques in Soil Stabilization for Road Construction

Jasmina Novakovic

Arcler Press
4164 Lakeshore Road
Burlington, ON L7L 1A4
Canada
www.arclerpress.com
Email: orders@arclereducation.com

ISBN: 978-1-77956-435-1 (Hardcover)

ABOUT THE AUTHOR

Jasmina Novaković is an accomplished physicist and educator with a rich background in applied and computer physics. Holding a bachelor's degree in Applied and Computer Physics, Jasmina has dedicated her career to making complex scientific concepts accessible and engaging. With years of experience as a physics teacher, she has a proven track record of inspiring curiosity and fostering a deep understanding of the subject among her students. Her passion for physics extends beyond the classroom. Jasmina is a committed advocate for science education and has actively participated in popularizing physics through hands-on experiments and educational initiatives. Her work includes conducting interactive sessions and workshops aimed at demystifying physics for learners of all ages. In addition to her scientific expertise, Jasmina is a creative writer and translator fluent in English and Spanish, with native proficiency in Serbian. This linguistic skill set enhances her ability to communicate complex ideas and stories across cultures. This unique combination of skills positions her as a versatile professional capable of addressing a wide range of projects with precision and cultural sensitivity. Jasmina is also committed to community service, having volunteered in education, disaster relief, and legal aid. Her diverse background in physics, writing, and community support provides a unique and enriched perspective in all her endeavors.

TABLE OF CONTENTS

LIST OF FIGURES

LIST OF TABLES

PREFACE

The importance of reliable, durable, and cost-effective road infrastructure cannot be overstated, particularly in regions experiencing rapid urbanization and economic growth. Roads are the arteries of modern civilization, connecting communities, facilitating trade, and fostering economic development. Yet, one of the most persistent challenges in road construction remains the stabilization of soil, a fundamental component that determines the longevity and performance of road networks. The properties of soil vary dramatically from one location to another, making it necessary for engineers to adopt advanced techniques to stabilize soils effectively before they can support the weight and stresses of modern traffic loads.

"*Advanced Techniques in Soil Stabilization for Road Construction*" provides a comprehensive exploration of the various methods and materials available for stabilizing soils in road construction projects. This book is aimed at civil engineers, geotechnical professionals, road construction managers, and students who wish to deepen their understanding of soil stabilization. It covers both traditional and emerging techniques, reflecting the diverse approaches that can be applied depending on specific soil types, environmental considerations, and project requirements.

The book is structured into 10 chapters to create a holistic understanding of the subject.

Chapter 1 serves as a foundation by defining the concept of soil stabilization, its relevance in road construction, and the various challenges posed by weak soils. It provides an overview of the techniques that will be discussed in later chapters.

Chapter 2 emphasizes the importance of understanding the physical and mechanical properties of soil before selecting appropriate stabilization methods. The chapter includes a discussion on various classification systems, as well as laboratory and field tests used to assess soil suitability for stabilization.

Chapter 3 explores the use of chemical additives, such as lime, cement, and fly ash, which are commonly used to improve the load-bearing capacity of soils. The chapter details how these materials interact with different soil types and the advantages of each chemical method.

Chapter 4 examines the methods like compaction, blending, and other physical techniques for improving soil strength without altering its chemical composition. This chapter also examines how mechanical stabilization methods can be combined with chemical techniques for optimal results.

Chapter 5 reflects the growing trend of incorporating sustainable practices in road construction. The chapter highlights how materials like crushed concrete, asphalt, and industrial by-products can be used as stabilizers, offering both environmental and economic benefits.

Chapter 6 introduces biological methods that utilize microorganisms to improve soil properties. Though relatively new, bio-stabilization is gaining attention as a sustainable and eco-friendly alternative to traditional methods.

Chapter 7 presents case studies and research on innovative uses of natural and waste materials to stabilize problematic soils. The chapter discusses the efficacy and potential environmental benefits of these unconventional materials.

Chapter 8 addresses the environmental considerations associated with various stabilization methods. The chapter also examines the life-cycle analysis of soil stabilization projects, emphasizing the importance of minimizing the environmental footprint of road construction activities.

Chapter 9 focuses on the practical aspects of planning, designing, and executing soil stabilization projects. It covers everything from site investigation and material selection to construction techniques and project management strategies.

Chapter 10 discusses the importance of ongoing monitoring and testing to ensure that stabilized soils continue to meet the performance requirements of modern roadways. It emphasizes quality control measures and the role of new technologies in real-time monitoring.

This book aims to be a valuable resource for practitioners and researchers alike. Providing detailed insights into advanced stabilization techniques and their applications, it equips readers with the knowledge to design, implement, and manage road construction projects that are both durable and sustainable.

—Author

CHAPTER 1

SOIL STABILIZATION IN ROAD CONSTRUCTION

Contents

INTRODUCTION

Enhancing the shear strength parameters of soil to increase its bearing capacity for road construction is known as soil stabilization. When the soil that can be used for construction is unsuitable to support structural loads, it is necessary. If soils aren't treated to improve their physical characteristics, they typically show undesirable engineering properties. Stabilization can improve a soil's shrink-swell characteristics and shear strength, which increases a subgrade's ability to support pavement and its foundations by enhancing its load-bearing capacity. The purpose of soil stabilization is to improve the shear strength of the soil mass and decrease its permeability and compressibility in earth structures. Stabilization is achieved by adding additives to the reaction mechanism, which reduces and maintains the soil moisture content while also positively influencing the mechanism's strength. Thus, these methods of soil stabilization can be achieved in a variety of ways and are recommended for the majority of construction systems. Mechanical stabilization and chemical stabilization are the two main categories into which all of these techniques can be divided. By altering the soil's gradation, mechanical stabilization improves its characteristics; chemical stabilization involves adding a physico-synthetic material to the soil, which combines with the clay particles to fill in the voids and reduce the amount of water required to maintain a stable mix and, ultimately, a stable framework.

1.1. IMPORTANCE OF SOIL STABILIZATION FOR ROADS

Numerous soil types are encountered during road construction projects, ranging from weak, soft soils to extremely wet soils, which can present serious difficulties. Clayey soils, for example, are soft soils with low load-bearing capacity that can settle and cause surface deformations on roads. By improving the properties of the soil and giving the road a solid foundation, soil stabilization helps to mitigate these problems. Additionally, soil stabilization strengthens the road's structural integrity, lowering the possibility of early failures and expensive repairs. Road surfaces can withstand the damaging effects of traffic loads, freezing and thawing, and wetting and drying better when subbase soils are stabilized. As a result, roads that experience soil stabilization have increased resilience and a longer lifespan.

An essential component of building roads is stabilizing the soil, which gives the structure a strong base and improves the longevity and functionality of the roads. Soil stabilization ensures the longevity and structural integrity of roads by enhancing the engineering properties of soil and reducing problems like erosion, deformation, and settlement. For a variety of soil types and project requirements, different soil stabilization techniques, such as mechanical, chemical, and cement stabilization, provide efficient solutions. These techniques enhance dust control, moisture regulation, durability, workability, and bearing capacity. Adequate preparation, mixing, compaction, curing, and comprehensive soil testing are crucial when choosing and implementing soil stabilization techniques. The efficacy and caliber of soil stabilization are guaranteed by conforming to acceptance criteria, which include compressive strength, moisture content, density, and surface smoothness. Almost all road construction projects as well as those involving the stabilization and repair of asphalt roads require the use of earthwork. All roads and highways are built on soil, whether it is in the form of transported and reworked base, undisturbed subgrade materials, or embankment material.

Due to its low strength and stiffness, clay liquefies and softens when exposed to large amounts of water. This characteristic makes construction challenging. Weak soil can compromise land stabilization and building foundations while causing fissures in buildings and asphalt pavement road surfaces, which has resulted in serious geotechnical engineering problems. It is necessary to shift the loads from the pavement to the base and subbase. The road will deteriorate rapidly in the absence of sufficient assistance. A strong foundation and soil stabilization techniques are necessary for a good road, and these require material stability. A material is stable whether it is wet or dry if it resists deformation under repeated or prolonged stresses and experiences little to no volume change.

1.1.1. Road Stabilization Benefits

Soil stabilization offers numerous benefits for road construction projects. Let's explore some of the key advantages:

1. **Increased Bearing Capacity:** Stabilization boosts the subgrade soil's bearing capacity by improving the soil's engineering characteristics. This prevents undue settlement or deformation of the road and enables it to accommodate higher traffic volumes.
2. **Improved Durability:** Stabilized soils are more resilient to weather and traffic-related deterioration. As a result, the road has a longer service life, which lowers maintenance expenses and inconveniences.
3. **Improved Workability:** Highly plastic, sticky soils are challenging to compact and place. The process of lime stabilization significantly reduces or eliminates any potential for swelling while also permanently changing the structure of clay minerals, making the soil less plastic and more friable.
4. **Moisture Control:** Reduced stability and softening of the soil

can result from an excessive moisture content. The process of stabilization, especially lime stabilization, aids in regulating the moisture content within the ideal range so that the soil retains its strength and stability even when it is wet.

5. **Minimize Subsequent Layers:** Pavement design can take the strength of soil stabilization into consideration. The stabilized layer's strength can minimize the need for additional pavement layers, saving resources, money, and time.

1.1.2. Soil Stabilization Methods

Depending on the particular project requirements and soil conditions, different soil stabilization techniques are used during road construction. Here are some commonly used techniques (Figure 1.1):

1. **Mechanical Stabilization:** With this technique, undesirable soil is removed and granular materials, like gravel or crushed stones, are added to or used as replacements for the original soil. The granular material strengthens and stabilizes the soil, increasing its ability to support weight. This method works especially well in remote locations or when there is easy access to rock on the property.

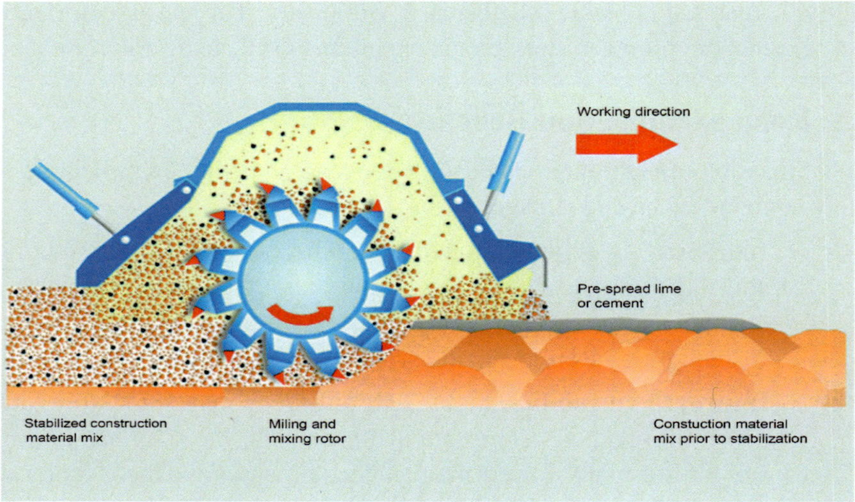

Figure 1.1. *Overview of soil stabilization methods in road construction.*

Source: https://media.springernature.com/lw685/springer-static/image/chp%3A10.1007%2F978-3-030-01911-2_3/MediaObjects/473997_1_En_3_Fig1_HTML.png.

2. **Cement Stabilization:** A common technique is soil cement stabilization, which uses cement to bind more granular soil particles together, resulting in a compacted layer that is stronger and more stable. This method works especially well on soils with more sand.
3. **Lime Stabilization:** Lime stabilization is modifying the soil's characteristics by adding chemicals based on lime. For instance, quicklime is frequently used to lessen the clay particles' plasticity, strengthen the soil, and make it easier to work with, get rid of potential swell, and speed up drying. This method works especially well on soils that contain clay.

1.1.3. Soil Stabilization Selection Guidelines

Many factors need to be taken into account when choosing the best soil stabilization technique, including the type of soil, moisture content, traffic volumes, surrounding conditions, and project requirements. For soil stabilization in clayey soils, lime-based products like quicklime and lime kiln dust are far more recommended than Portland cement. The following guidelines can assist in the selection process:

1. **Soil Testing:** To evaluate the traits and qualities of the clayey soil, carry out extensive soil testing. The best stabilization technique will be determined with the help of this information. In particular, tests like the California Bearing Ratio (CBR) test, the Atterberg limits test, and compaction tests can shed light on the behavior of the soil and help calculate the right quantity of lime needed for stabilization.
2. **Chemical Selection:** Select the suitable stabilizing product. Because of their high reactivity and efficiency in changing the clay particles, quicklime (calcium oxide) and lime kiln dust (a byproduct of lime production) are frequently utilized in clayey soils. Seek advice from Mintek Resources to ascertain the best lime product and dosage for the particular soil circumstances.
3. **Mixing and Application:** Using specialized machinery—typically a soil reclaimer—the stabilization product should be thoroughly mixed with the soil after it has been chosen. The stabilizing agent and soil particles are properly distributed and interact thanks to the mixing process. To ensure that lime or cement is activated (hydrated) and to reach the appropriate soil moisture content, adequate moisture control must be maintained during the mixing process.
4. **Compaction:** Compaction is required to attain the required density and strength of the stabilized soil after mixing and reaching the desired

mellow, which usually takes less than four hours. It is important to apply the required pressure and achieve uniform compaction throughout the stabilized layer using compaction equipment, such as rollers. The stability and load-bearing capacity of the road are guaranteed by proper compaction.

5. **Curing:** Give the stabilized soil time to heal and get stronger. To prevent excessive drying, the stabilized layer must be kept moist during the curing process by being watered or covered with geotextiles. For the stabilized soil to form stable bonds and create a strong and long-lasting stabilized soil layer, there must be a sufficient amount of curing time.

1.1.4. Soil Stabilization Acceptance Criteria

To ensure the effectiveness of soil stabilization, certain acceptance criteria should be met. These criteria may include:

1. **Compressive Strength:** Through laboratory testing, the stabilized soil should reach a predetermined compressive strength. This guarantees that the layer that has stabilized can tolerate the anticipated traffic volumes and additional environmental strains. The curing period, which can end as soon as compaction is completed or take up to 48 hours, depends on many variables. Generally speaking, the curing period is defined as the amount of time required to reach the minimum desired strength.
2. **Moisture Content:** The ideal moisture content range for the soil should be reached by the stabilized soil. This guarantees the stabilized layer's performance and stability in a range of moisture conditions.
3. **Density:** For sufficient support and load-bearing capacity, the stabilized soil layer that has been compacted to the appropriate density must be reached. Density testing can confirm that the necessary density has been reached. Examples of these tests are the Proctor compaction test and proof roll.
4. **Surface Smoothness:** The stabilized road surface should be smooth and even, ensuring a comfortable and safe driving experience for users as well as slightly graded to shed water. Regular inspection and measurement of the road surface can help identify any unevenness or deviations from the required smoothness.

1.1.5. Subgrade Stabilization

Subgrade stabilization is the process of providing constructability and access over extremely soft soils through the use of reinforcement components. Subgrade solutions produce a level playing field with unwavering assistance. The process of preserving or altering one or more soil features to enhance the soil's properties and performance from an engineering standpoint is known as soil stabilization. The process of altering a natural soil's properties by mixing in specific soil, cement, or other chemicals is known as "soil stabilization." The soil can be stabilized using several techniques to change its characteristics and enhance its engineering performance. One way to stabilize soil is to add stabilizing material to an undisturbed soil deposit and let it seep through soil voids to interact, or you can mechanically combine natural soil and stabilizing material to create a homogeneous mix. When less appealing road soils are stabilized with additives, their properties are enhanced. When applied correctly, these soil stabilizers can enhance and maintain the moisture content of the soil, encourage cohesion between soil particles, and function as cementing and water-proofing agents.

1.1.5.1. Subgrade Stabilization with Cement

Particles of inorganic matter that have been finely ground are called cement. When cement and water are combined, the cement paste sets and hardens due to hydration processes and reactions, keeping its stability and strength even under water. Portland cement, or regular cement, is the most widely used type. Raw materials for its production include iron oxide, silica, lime, and alumina. Before burning, these materials are mixed and crushed in the appropriate amounts in a rotary kiln. Grind clinker and gypsum to a fine powder, then cool. This is how cement is made. Cement typically contains two main chemical constituents: aluminates and calcium silicates. In the construction of roads and earth dams, cement is a commonly used material for stabilizing the soil. Clayey or sandy soils can benefit from its application to maintain stability. Similar to lime, cement improves clayey soil workability and raises the plasticity index while lowering the liquid limit.

Cement stabilization is also known as soil-cement stabilization. Soil-cement is a highly dense mixture of wet soil, cement, and water that is packed tightly. Once the soil-cement mixture cures, a sturdy, rigid base course is formed. A thin, light- or medium-traffic wearing surface can be stabilized with soil cement as a base asphalt, or it can be used as support for a high-type flexible or stiff pavement. A 3–16% volume percentage of the compacted mixture is added to the native soil as an amendment using Portland cement. It can stabilize almost any soil type; however, silts and heavy clays need a higher cement percentage for the stabilization process to be effective. In contrast to other stabilization techniques, cement stabilization gives the soil material structural strength by hardening it, as opposed to cohesiveness or waterproofing. For soil-cement, the three main control factors are appropriate cement content, acceptable compaction, and appropriate moisture content. These should be ascertained by laboratory testing of representative soil samples. To achieve maximum compaction, pulverized dirt is mixed with a small amount of water and Portland cement. The cost fluctuates greatly.

1.1.5.2. Lime Subgrade Stabilization

Soils rich in clay tend to become more pliable when their moisture content rises. The soil can be stabilized with quicklime or hydrated lime to improve plastic clay soils. With clay, lime, a potent alkaline base, reacts chemically to produce complex silicates and other cementing agents. Although powdered limestone is not advised owing to its chemical inertness, high calcium or dolomitic lime can be used. Marginal and submarginal soils can be improved into suitable base and subbase materials with the help of lime. Moreover, it improves the workability and pulverization of polymeric soils. Building can proceed more quickly thanks to lime's tendency to waterproof the soil to some extent and accelerate its drying out after saturation. To figure out how much lime to use, lab testing is done. Subgrade therapy, on the other hand, typically uses a concentration of 3 to 6%.

Lime is then applied, usually using a mechanical spreader, after the subgrade soil has been scarified and ground to a depth of six inches. The moisture content is raised to 5% or more above the optimal level by adding enough water. The mixture of lime and soil takes one to seven days to cure. Up until 60% of the material passes through a number four sieve and all of the material passes through an inch sieve, the mixing and crushing procedures are repeated. To compact the material, vibrator compactors or pneumatic rollers are utilized. The compressed layer needs three to seven days to cure before the next layer is applied. It is best to avoid heavy traffic on the road while the road dries out.

1.1.5.3. Calcium Chloride

Calcium chloride is used to maximize the performance of appropriately graded quality aggregates; it is not used to modify the properties of inferior or dubious materials. Moisture can be drawn and retained from the air by calcium chloride. Reducing the rate at which moisture evaporates from the mixture during compaction helps to retain moisture throughout the soil-aggregate combination's service life and expedites the compaction process. In many cases, adding small amounts of calcium chloride will increase density for a given compaction effort. About 12 lbs is the initial application rate per sq. yd. per thickness of an inch.

1.1.6. Purpose of Soil Stabilization

It is not appropriate to build on every building site. A contractor at such a site will usually have them. One should implement soil stabilization for several reasons, such as:

Chemical and mechanical soil stabilization:

- Substituting poor-quality soils with aggregates with better engineering qualities.
- Strengthening of the soil and, as a result, its bearing capacity.
- Dust control is essential for a healthy working environment.
- Waterproofing is used to keep natural or man-made structures from deteriorating.
- Encourage the use of waste geomaterials in construction projects.
- Finally, improving the soil qualities on-site.

1.1.6.1. How Do You Stabilize Soil?

Physical and/or chemical techniques are used in soil stabilization to prepare the soil for construction. During soil stabilization, an existing surface is ground up using a soil stabilizer. Subsequently, bituminous or chemical additives are used to improve characteristics and increase bearing capacity. Examples of such additions are cement, fly ash, lime, and soil cement. The enhanced material is then graded and compacted to produce a stable foundation that can withstand the necessary stresses.

The following are the main steps in soil stabilization:

- Evaluation of soil qualities
- Selection of a method for restoring missing properties
- Creation of a stabilized soil mixture for the required purpose
- Calculation of compaction parameters

What are the soil stabilization methods?

- **Mechanical Stabilization** — by mixing and compacting two or more soils and/or aggregates, the goal is to achieve dense, well-graded material.
- **Chemical Stabilization** – is the process of modifying the chemical make-up of soil by adding different additions such as lime, cement,

or fly ash, or by adding compounds such as polymers, resins, and enzymes.

- Planting of vegetative cover to reduce wind, water, and soil erosion is referred to as biological stabilization. Although various techniques of stabilization should be utilized to assist the growth of seeds and seedlings in the beginning, the roots hold and agglomerate soil particles together.

To manage rivers, and build airfields, parking lots, landfills, embankments, existing pavement roads and foundations, agriculture, and mining sites, among other projects, stabilizing soils is a common practice. Depending on the situation, either a single technique or a combination of the two may be used for stabilization. The two types of stabilization techniques that these methods fall into are chemical stabilization and mechanical stabilization. The process of improving the properties of soil by changing its gradation is known as mechanical stabilization. Combining various soil types is the process of mechanical soil stabilization, which improves stability. Some soils have poor engineering properties. These soils are characterized by high concentrations of silt, clay, or fine sand. Consequently, the overall qualities of *in-situ* soils can be enhanced by adding more suitable soils containing granular material. The mechanical strength of the aggregate, the mineral content of the constituents, the gradation of the mixture, the soil's flexibility as the binder, and the degree of compaction all contribute to the stability of soil-aggregate mixtures. Regarding mineral composition, while chlorides and carbonates aid in stabilizing the soil, sodium sulfate and carbonate have a detrimental effect because of the large volume changes they cause when the water content of the soil varies.

Mechanical soil stabilization has the following drawbacks:

- large cost and carbon footprint,
- heavy equipment, labor, and time demands, and substantial traffic disruption

The two most often used chemical additions are cement and lime. A structure's strength and longevity grow as its cement content does. The use of cement is limited because of its high cost. The presence of magnesium sulfate, which reacts with hydrated cement to produce a drop in strength, may result in the durability of cement-treated soil being reduced by the influence of surface chemical variables, organic matter, and sulfate content. Applying lime results in dust, which makes it inappropriate for areas with high population density. Correct water content is essential for ideal compaction. Lime adds to the cost of the process because it requires more time to mix.

1.1.7. Materials Used in Soil Stabilization

The materials for stabilizing soil differ based on the method. The following is a list of mechanical, chemical, and biological soil stabilization techniques:

- Different grades of soil.
- Different grades of aggregates.
- Seedlings
- Seeds
- Hydromulch mixtures
- Hydroseeding mixtures
- Geomaterials – geogrids, geoblankets
- Polymers -synthetic and natural
- Synthetic resins
- Emulsions
- Cement
- Lime
- Fly ash
- Bitumen
- Recycled and waste products – solid municipal, mining and industrial wastes.

In pavement construction, the most common techniques for stabilizing clay soil are cement and lime stabilization. Lime, also called calcium carbonate, is the most ancient chemical stabilizer still in use today for stabilizing soil (Al Tabbaa and Stegemann, 2005; Kowalski and Starry, 2007).

1.2. THE FUTURE OF ROAD CONSTRUCTION: EXPLORING SOIL STABILIZATION TECHNIQUES AND BENEFITS

To ensure the best possible quality, longevity, and resilience of road infrastructure, soil stabilization is an essential component of road construction. Soil stabilization gives pavement layers stable support by improving the physical characteristics of soil, such as strength, stiffness, and permeability. Roadway performance and safety are ultimately improved by this process, which reduces the likelihood of rutting, cracking, and other pavement distress. The many soil stabilization methods that are frequently used in road construction will be discussed in this section, along with their uses and advantages.

There are two main categories of soil stabilization techniques: mechanical and chemical. While chemical stabilization entails the addition of stabilizing agents like lime, cement, or polymers, mechanical methods physically modify the soil through compaction, mixing, or reinforcement. One can choose the best strategy for a given road construction project by being aware of the benefits and constraints of each technique. To provide a stable foundation for roads, soil stabilization is the process of enhancing the engineering characteristics of soil, such as strength, permeability, and resistance to weathering. Stabilizing techniques can be broadly divided into two groups: mechanical and chemical. They differ based on the particular requirements of each project (Al Tabbaa and Stegemann, 2005; Kowalski and Starry, 2007) (Figure 1.2).

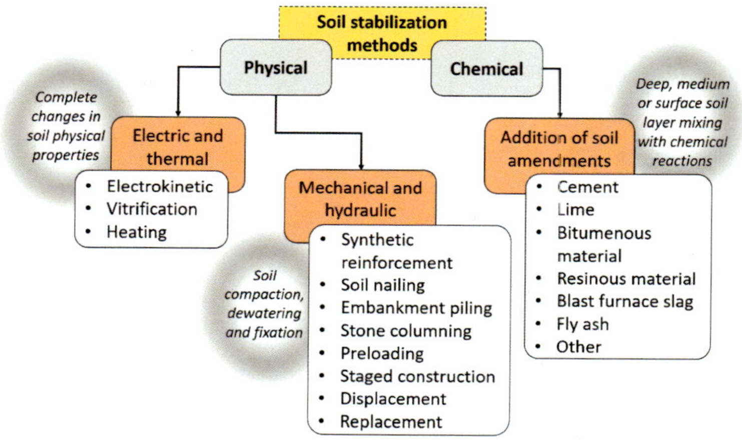

Figure 1.2. *Complete guide to soil stabilization and its benefits.*

Source: Vincevica-Gaile Z, Teppand T, Kriipsalu M, Krievans M, Jani Y, Klavins M, Hendroko Setyobudi R, Grinfelde I, Rudovica V, Tamm T, et al. (2021). Towards sustainable soil stabilization in Peatlands: secondary raw materials as an alternative. Sustainability, 13(12):6726. https://doi.org/10.3390/su13126726.

1.2.1. Approaches of Mechanical Soil Stabilization

Through procedures like compaction, blending, and reinforcement, soil properties are physically altered as part of mechanical stabilization techniques. Typical mechanical stabilization techniques consist of the following:

- **Compaction:** With this technique, air voids are reduced and density and strength are increased by applying pressure to the soil using large machinery like rollers. Because compaction increases bearing capacity and decreases settlement, it is necessary to create a subgrade that is suitable for road construction.

- **Blending:** Blending is the process of combining soil with additional materials to make a more uniform and stable foundation, such as aggregate or recycled pavement materials. This method works especially well for dealing with uneven soil conditions and for strengthening weaker soils with more durable materials.
- **Geosynthetics:** This technique improves bearing capacity and load distribution by incorporating synthetic materials like geotextiles and geogrids into the soil to reinforce and add extra support to troublesome areas.

1.2.2. Approaches of Chemical Soil Stabilization

Chemical stabilization techniques improve the engineering properties of the soil by adding stabilizing agents or additives. The following are a few typical chemical techniques for stabilizing soil.

- **Lime Stabilization:** Since lime and soil minerals combine to form cementation compounds, which strengthen, decrease plasticity, and improve moisture resistance in the soil, lime is frequently used to stabilize clayey soils. In road construction projects, base course and subgrade improvements are especially beneficial for lime stabilization.
- **Cement Stabilization:** Another well-liked stabilizing agent is cement, which when combined with soil particles and water creates cementation compounds that strengthen and last longer. To improve a variety of soil types, cement stabilization is frequently used in infrastructure projects like building new roads.
- **Polymer Stabilization:** Using this method, the soil is supplemented with polymers, which strengthen the soil's cohesiveness and increase its resistance to erosion and water. When it comes to projects that prioritize environmental concerns like reducing runoff and sedimentation, polymer stabilization can be especially helpful.

1.2.3. Comparison of Mechanical and Chemical Soil Stabilization

The best soil stabilization technique for a given project will depend on several variables, such as the type of soil, the demands of the project, and the available funds. A quick comparison of chemical and mechanical stabilization methods is given below:

- **Soil Compatibility:** Chemical techniques work better on certain types of soil, like silty or clayey soils, but mechanical techniques are more adaptable and can be used on a wider range of soil types.
- **Environmental Impacts:** Given the possibility of chemical additives seeping into adjacent water sources, chemical stabilization techniques should be more environmentally conscious. Nonetheless, these risks are being reduced by new developments in environmentally friendly stabilizing agents.
- **Cost:** Depending on the size of the project, the availability of materials, and the amount of labor needed, the cost of various stabilization techniques can differ considerably. Chemical stabilization is generally more economical in some circumstances, such as when major improvements are needed to the current soil conditions.

1.2.4. Case Study: Soil Stabilization Solutions with Terra-Firma Stabilization and Reclamation

As an example, a company says "When you work with Terra-Firma Stabilization and Reclamation, you can be sure that your needs for soil stabilization will be satisfied with the best possible knowledge and effectiveness. Our qualified staff provide:

- **Comprehensive Services:** To meet the needs of various project requirements, we offer a broad range of specialized soil stabilization and reclamation services, including mechanical and chemical stabilization methods.
- **Customized Solutions:** Our skilled staff collaborates directly with clients to create specialized solutions that meet the unique demands and difficulties of each project.
- **State-of-the-Art Equipment:** We utilize advanced equipment and technologies to ensure optimal results in each phase of the stabilization process.
- **Quality Assurance:** Our enormous experience and outstanding track record in the industry support our dedication to quality and compliance with industry standards."

1.3. COMPARATIVE ANALYSIS OF A THREE-PRONG APPROACH

The naturally occurring loose surface material of the earth that, when dug up or plowed, supports plant growth is known as soil. Another naturally occurring substance is clay, which is primarily made up of fine-grained materials—tiny particles with adhesive and plastic qualities. Because of its tiny holes and pores, clay can hold onto water, and water tends to cause expansion and shrinkage, which can result in settlement. Clay's tendency to soften and liquefy when exposed to large amounts of water makes it difficult to construct because of its low strength and stiffness. This has created significant difficulties for geotechnical engineers since poor soil can lead to cracks in pavement, damage to building foundations, and other issues. The modification or preservation of one or more soil properties to enhance the soil's performance and engineering qualities is known as soil stabilization.

In general, the process of improving one or more of a natural soil's properties by adding a specific soil, cementing material, or other chemical materials is referred to as soil stabilization. Therefore, stabilization examines the different techniques used to alter a soil's characteristics to enhance its engineering performance. Soil stabilization can be accomplished in two ways: either by adding stabilizing material to an undisturbed soil deposit and allowing it to permeate through soil voids to create an interaction, or by mechanically mixing natural soil or stabilizing material together to create a homogeneous mix. The characteristics of less ideal road soils are enhanced when soil is stabilized through the application of additives. These stabilizing agents can be used as cementing and water-proofing agents, to maintain and improve the moisture content of the soil, and to increase the cohesiveness of the soil particles.

When the subgrade is discovered to be clay soil, it typically presents a challenging issue in civil engineering projects since the foundation of any structure, including roads, is a crucial component in terms of load transfer to the earth. When moisture content in high clay soils is allowed to rise, the soils have a propensity to swell. Cement and lime stabilization are the most widely used techniques for stabilizing clay soils during pavement construction. The earliest known conventional chemical stabilizer for stabilizing soil is lime, also known as calcium carbonate. By contrasting the application of cement, brick dust, and sodium chloride (NaCl) in varying percentages with clay soil obtained from a borrow pit in a university setting in Ota, Ogun state, the work presented in this paper adds to the use of chemical stabilization technique. There are four different stabilizer concentrations by weight: 2%, 6%, 10%, and 14%. The open market purchases of cement, brick dust, and NaCl came from Abeokuta.

The fact that Nigerian roads are infamous for having potholes, longitudinal cracks, and numerous other pavement flaws that are well-known to those involved in road construction serves to highlight the difficulty. Potholes are ubiquitous in Nigeria and have significantly impacted both the country's high accident rate and decline in economic development. You can hardly go a mile without encountering one. All constructed roads have a specified design life, but most of the time they fail well in advance of that time. Some fail right after construction, while others fail after significant flooding. Some roads only last up to half of their life cycle, while others, with proper maintenance, last for their entire expected life. The rising water content of granular subgrade materials is one of the main reasons for these failures.

This study aims to determine the most effective additive among cement, sodium chloride, or brick dust to improve the geotechnical properties of clay soil. Soil stabilization is broadly defined as any method used to improve the soil's ability to carry load and CBR values of soil and thus render it able to resist the loads and adverse effects of construction activities. Utilizing waste materials or industrial production byproducts is becoming more common because it lowers carbon emissions.

This supports the use of brick dusts as a substitute for salt and cement in soil stabilization when it is recognized that they help reduce the amount of time needed for complete excavation and replacement, which was previously the case in cases of weak soils. It is imperative to conduct prescribed tests on a soil sample to understand the nature of the soil before making decisions about stabilizing it for engineering purposes (Al Tabbaa ad Stegemann, 2005; Kowalski and Starry, 2007).

1.3.1. Soil

According to one definition, soil is made up of a variety of relatively small particles that have been broken down from parent rock by chemical and mechanical weathering processes such as oxidation and carbonation, burrowing by plants and animals, temperature fluctuations, freeze-thaw cycles, and air or abrasion. Another way to think of soil is as the solid material that forms on Earth's surface as a result of biological activity and weathering on underlying hard rock or parent material.

Different people define soil according to what they need. For an engineer, it's a substance that can be used for building roads, runways, embankments, bridges, tunnels, culverts, basements, and supporting structures like retaining walls and quays. For a geologist, it's the result of historical surface processes; for a pedologist, it's current physical and chemical processes. Sabat (2012) also pointed out that an engineer can enhance the engineering qualities of soil to make it fit for building using a variety of methods. These include stabilization with waste materials like dust or powder both with and without a binder like cement, lime, etc. Other well-known dust- or powder-like waste materials that have been effectively used for stabilizing expansive soil include marble dust, brick powder, pyroclastic dust, baryte powder, and quarry dust.

1.3.2. Clay Soil

Of all the soil types, clay has the smallest particles—each one so tiny that it can only be seen under an electron microscope. As a result, there are many clay particles in a comparatively small amount of space instead of the spaces that would typically exist between larger soil particles. Clay-heavy soils typically have very dense soil structures due to the small particle sizes of clay soils.

1.3.3. Clay Mineralogy

The term "clay" describes a naturally occurring substance that is primarily made up of fine-grained minerals. It is typically plastic when it contains the right amount of water and hardens when it is fired or dried. Generally speaking, the minerals in clay are silicates that are smaller than 2 microns, or one-millionth of a meter, or roughly the size of a virus. Since clay is so prevalent at the earth's surface, it frequently forms shales and is a significant component of almost all sedimentary rocks. Clay minerals have special properties such as cation exchange capabilities, plastic behavior when wet, catalytic abilities, swelling behavior, and low permeability due to their unique crystal structures and small particle sizes. The hydrous silicates that make up clay minerals are primarily composed of iron, magnesium, and aluminum. When heated, they lose their constitutional and adsorbed water, producing refractory material at high temperatures. The affinity of the clay surface for water, which results from a net negative charge on the surface of a clay particle that causes it to adsorb water and other fluids, is primarily responsible for the plasticity of clay minerals.

Thus, clay minerals, even in trace amounts, can have a significant impact on the engineering behavior of a soil sample. A soil sample's clay fraction will have an increasing effect on the sample's behavior as its clay content rises. The addition of bentonite to a granular soil sample demonstrates the significant impact of clay minerals on soil behavior. Sodium montmorillonite is the main component of the clay mineral bentonite.

1.3.4. Cement

Cement is defined as an inorganic material that has been finely ground. When combined with water, it forms a paste that undergoes a hydration reaction to set and harden, and it maintains its strength and stability even when submerged in water. The most widely used kind of Portland cement is regular grade. It is made of iron oxide, silica, alumina, and lime as raw materials. In a rotary kiln, these ingredients are ground, mixed, and burned in the proper ratios. To make cement, the clinker is cooled, combined with gypsum, and ground into a fine powder. The primary chemical constituents of cement are aluminates and calcium silicates. Cement is a widely used soil stabilizing material, especially in the building of earth dams and highways. Both sandy and clayey soils can be stabilized with it. Similar to how lime works, cement also reduces the liquid limit and the plasticity index while raising the workability of clayey soils.

1.3.5. Salt

The salinity of the soil water can influence the physical characteristics of the soil by causing fine particles to aggregate. This process, called flocculation, is advantageous for changing the soil's necessary engineering properties. The use of sodium chloride is recommended due to the abundance of salt and economic viability. Salinity affects soils in the opposite way that sodium does. When there are too many large sodium ions between two clay particles, the forces holding them together are broken. The clay particles swell and disperse the soil as a result of this separation. According to a 2009 study by Işık and Berrin, gypsum performs as an additive for treating clay soils by increasing their swell potential and strength. It has been investigated how adding chloride salts, such as NaCl, $MgCl_2$, and $CaCl_2$, affects the silty clay soil's engineering characteristics. The findings demonstrated that while the optimal water content decreased, the maximum dry density (MDD) increased as the percentage of each chloride salt increased. With an increase in salt content, it is known that the plastic limit, liquid limit, and plasticity index will decrease. Additionally, the results demonstrated that an increase in salt content was accompanied by an increase in unconfined compressive strength (UCS) (Al Tabbaa and Stegemann, 2005; Kowalski and Starry, 2007).

1.3.6. Brick Dust

It is well known that brick dust, particularly when it is burnt clay dust, increases the amounts of alumina and amorphous silica in soil. Because silica and alumina are present, when clay bricks are fired at a temperature lower than 950°C and ground into a fine powder, they react pozzolanically when combined with lime. But it's important to remember that raw clay has no pozzolanic value by itself. Although the exact composition varies depending on the source, in general clay should contain 20–30% alumina, 50–60% silica, and the remaining material should be composed of iron oxide, calcium carbonate, and magnesium carbonate. For instance, Bhavsar et al. (2014) conducted a study on the application of industrial waste, such as brick dust, in soil stabilization. The study's benefits included the appropriate disposal of this waste, preservation of biodiversity, enhancement of soil properties like strength, and reduction of permeability, preservation of the soil's natural state and the planning and building of economic structures. The application of brick dust to black cotton soil yields good results, as increasing the amount of brick dust lowers the soil's swelling index and stabilizes it.

Industrial clay bricks are crushed into dust particles to create brick dust, a material that is readily available in the area. Even though several businesses in Lagos make them into building bricks, this admixture is offered and sold on the open market in Nigeria, which includes Ota. As a result of the soil's decreased plasticity as a result of the reduction in clay content of the soil due to the replacement of clay with brick dust, the work by Kumar et al. (2016) also reported a decrease in the swelling index of clay soils when mixed with lime and brick dust.

1.3.7. Advantages of Soil Stabilized Roads

When building new roads, soil stabilization is a common practice. To improve their physical characteristics, soils need to be stabilized because they have unfavorable engineering characteristics. Structured soil layers that can support traffic loads condition soil-stabilized roads. These roads can support foundations and pavements with greater load-bearing capacity. Building roads with soil stabilization has several advantages.

- **Increased Road Material Stability:** The weight of traffic and the heavy load of vehicles will cause roads to deteriorate rapidly without proper support. Roads stabilized by soil offer material stability that doesn't deteriorate under heavy or continuous loads.
- **Reduced Road Maintenance Costs:** Soil stabilization reduces the amount of maintenance needed on roads compared to unsterilized ones. Potholes, rutting, and mudding—problems with unsterilized roads—do not occur on soil-stabilized roads. Money for road maintenance could be saved and put toward other projects if these maintenance expenses weren't incurred.
- **Saved Construction Costs:** Unstable road conditions frequently cause weather delays and setbacks for major construction projects. Companies lose money and time as a result of these unplanned delays. Rainfall setbacks are eliminated and the road surface remains stable for cars to drive on, so soil-stabilized roads are less likely to experience these delays. Additionally, soil-stabilized roads reduce the amount of time needed to dispose of subpar soil.
- **Dust Control:** Construction sites and unpaved roads are inherently problematic due to dust. The amount of dust produced by a road will vary depending on its location; nevertheless, roads in dry areas are more likely to produce dust. In addition to creating hazardous working

conditions for construction workers, dust can draw complaints from locals living in these areas. The treatment of soil-stabilized roads binds the soil particles, lowering dust and dust-related health issues.

- **Decreased Erosion Problems:** Erosion can have negative effects on a building project. Most building sites run the risk of experiencing issues related to erosion. Roads with soil stabilization lessen the chance of erosion's negative consequences. The process of stabilizing soil involves bonding soil particles together. There is essentially no chance of a road eroding or sloping because the soil sediment is sealed in place (Al Tabbaa and Stegemann, 2005; Kowalski and Starry, 2007).

1.4. THE PRIMARY NEED FOR SOIL STABILIZATION

The process of changing one or more soil properties mechanically or chemically to produce a better soil material with the required engineering properties is known as soil stabilization. When building a road, stabilizing the soil can increase the pavement's resilience and durability. Moreover, it can halt dust and erosion from occurring on the road's surface. Creating a soil material or soil system that will endure throughout the project's design lifespan under the project's intended use conditions is the main goal of soil stabilization. Similar to how soils differ globally, so too do their engineering characteristics. Thus, soil testing is crucial to the accomplishment of soil stabilization. The chosen method of stabilizing the soil should be tested in a lab before construction begins, ideally before choosing or purchasing materials. The most important part of a road is the highway pavement base. Road pavements are therefore vulnerable to the behavior of the soil that supports them. For road engineers, finding a way to reconcile tightening environmental regulations, limited budgets, and road performance is becoming more and more difficult. Treatments to improve the conventional pavements' long-term performance are becoming less and less cost-effective. Road budgets, particularly those for maintenance, seem to be declining annually about the work that needs to be done.

The long-term performance of pavement structures is often influenced by the stability of the underlying soil formations. The engineering design of these constructed facilities must comply with minimum standards for the structural integrity of each layer of pavement for the superimposed loads to be supported and distributed. These layers have to withstand excessive permanent deformation, shear, and severe deflections that can lead to fatigue cracking in the layers above. Since inexpensive natural earth materials don't always meet these requirements in their natural state, it might be necessary to enhance their engineering properties to convert them into valuable building materials. This is usually accomplished by chemically or physically stabilizing these troublesome soils. In-situ subgrades often lack the strength and stability required to guarantee satisfactory performance under traffic loading and environmental demands. Even though stabilization is a workable alternative for improving soil characteristics, the engineering properties that come from stabilization differ significantly because of the heterogeneity of geologic deposits, variations in soil composition, and variations in the physical and chemical interactions between the soil and potential stabilizers. Because of these variations, stabilization calls for the use of site-specific treatment options.

1.4.1. Procedures of Soil Stabilization

Soil stabilization can be accomplished through mechanical, chemical, electrical, or thermal methods. Seldom are the final two options selected. Compaction, also referred to as mechanical stabilization, is the process of densifying soil by applying mechanical energy. Densification occurs as air escapes from soil pores, largely without affecting the water content. In cohesionless soils, where compaction energy can cause particle interlocking and rearrangement, this method performs particularly well. Nevertheless, the method might not be effective if these soils exhibit significant variations in moisture. A decrease in the efficiency of compaction can also result from an increase in the fine content of the soil, or the percentage smaller than 75 μm. This is since cohesion and interparticle bonding inhibit particle rearrangement during compaction. Chemical stabilization, or the alteration of the physio-chemical properties of these fine-grained soils, is a more effective long-term stabilization strategy than compaction. Chemical stabilization of non-cohesive, coarse-grained soils—soils with more than 50% of their weight coarser than 75 μm—is also beneficial if a notable stabilization response can be obtained in these soils. In this case, the strength enhancement may be significantly greater—more than 10 times higher—than the strength of the untreated material. The most common methods of soil stabilization for roads include:

- Mechanical stabilization.
- Lime stabilization.
- Cement stabilization.
- Lime-Fly Ash (with or without cement) stabilization.
- Bituminous stabilization.
- Chemical stabilization.
- Geotextiles, fibers, prefabricated materials, etc.

1.4.2. Selection of Stabilizers

When selecting a stabilizer, it's crucial to take into account the type of soil that needs to be stabilized, how the stabilized layer will be used, the desired kind of soil improvement, the stabilized layer's strength and durability requirements, cost, and environmental factors. Depending on the texture, plasticity, or granularity of the soil, some general characteristics make some stabilizers better. For a given type of soil, more than one candidate stabilizer may be appropriate. For example, Portland cement can be applied to a range of soil types; however, it is best to avoid using more plastic materials because it is important to thoroughly blend the cement with the fines fraction (<0.075 mm). Well-graded granular materials with sufficient fines to form a homogeneous mixture, or floating aggregate matrix, are ideal for stabilizing Portland cement. Medium to high plasticity soils will become less plastic, more workable, less prone to swelling, and stronger when lime is added. By converting weak subgrade soils into a "working table" or subbase and mixing them with weak granular base materials, like clay-gravels, lime is used to stabilize a wide range of materials and produce a robust, superior base course.

Fly ash is almost always used with lime in soils that have little to no plastic particles because it is a pozzolanic material that reacts with lime. It is often found to be beneficial to add a small amount of Portland cement, lime, and fly ash to the mixture for added strength. Base course stabilization has been accomplished with success using lime, cement, and fly ash (LCF). For surfaces that are stronger and more weatherproof, bituminous compounds and asphalt are both utilized. Granular and silty sandy soils work well for stabilizing asphalt because it's important to cover every particle of soil. The optimal stabilizer selection may also be influenced by extreme weather, which may encourage the use of certain stabilizers while discouraging the use of others, regardless of cost. In general, particular attention is needed in hot, dry, and cold, rainy climates (Al Tabbaa and Stegemann, 2005; Kowalski and Starry, 2007).

1.4.3. Mechanical Stabilization

Mechanical stabilization is the process by which two naturally occurring forces inside the soil—internal friction and cohesiveness—develop. Compaction on its own can stabilize the soil in certain situations. Usually, the only way to stabilize the local soil is to add enough soil or gravel components. Mechanical stabilization is used when suitable grading and plasticity of locally available soil or gravel materials are unavailable. Mechanical stabilization involves the necessary amounts of mixing or blending of two or more selected materials to alter the plasticity and particle size distribution. On-site mixing is an option before final shape and compaction. A typical use for mechanical stabilization is mixing a fines-free granular material with sand-clay. Strength, compatibility, imperviousness, and abrasion resistance could all be enhanced by this material blending.

1.4.4. Lime Stabilization

The soil will change to provide maximum strength, and become less plastic, more workable, and less prone to swelling with lime stabilization. For every type of soil, there is a recommended amount of lime, and adding more will degrade the properties of the mixture. The amount of lime needed to stabilize a material (measured in percent by mass) depends on the type and quantity of clay minerals present in the soil. By decreasing the plasticity index, small quantities of lime (1 to 3%) can be used to stabilize some soils, such as clayey gravel with an acceptable grading but moderately high plasticity. The material constitution may change significantly when lime contents of 3 to 6% are used. Most plastic soil materials react well with lime in general, including clayey sands (SC) and salty clays (ML). It is possible that materials with a plasticity index of less than 10% won't react quickly. It is necessary to test the material's reactivity to lime to find out. Minor lime treatments should maintain some cohesiveness of poorly graded clayey sand and gravels in the stabilized soil. Failures will occur if too much is added because they will become friable (easily crumbled or pulverized) and completely non-cohesive.

Therefore, base material treated with lime should comply with the grading guidelines generally provided for untreated material. All fine-grained soils treated with lime typically exhibit decreased plasticity, improved workability, and decreased volume change. However, some soils lack characteristics that increase their strength. It is important to emphasize that a wide range of factors influence the characteristics of soil-lime mixes. The type of soil, the kind of lime, the amount of lime added, and the curing conditions (temperature, moisture content, and time) are the most important variables.

1.4.5. Cement Stabilization

The quality of the soil can be altered or improved, or it can be transformed into a cemented mass with greater strength and durability by using Portland cement. The amount of cement required will depend on whether stabilization or modification of the soil is required. An assortment of cement types have been successfully used to stabilize soils. Cement stabilization has been used extensively in pavement construction. On the other hand, cement is usually not a good stabilizing component for a pavement's wearing course. It is impossible for the cementation linkages to withstand traffic pressure if they are not covered by a wearing surface. Cement cannot be reworked after the initial mixing and subsequent setting, in contrast to lime. Additionally, reworking cement with maintenance instruments like graders is not feasible. However, it can also be used as a stabilizing agent for sub-bases.

Cement can be used to stabilize a wide range of soil types, including sandy materials and fine-grained silts and clays. Cement is usually used with clays or silts for fine-grained materials when the plasticity index (PI) is low. Cement stabilization should generally be avoided when sulfate content exceeds 1%, though local guidelines may allow higher thresholds.

The amount of cement required to alter the soil and improve its quality is determined through trial and error. To reduce the soil's PI, several samples of soil-cement mixtures at different treatment levels must be prepared. Next, each mixture's PI needs to be determined. The value needs to be adjusted to find the design cement content using the full sample weight given in the equation because it was calculated using that of the material's minus 40%. The least amount of cement needed to achieve the required PI is chosen.

$A = 100BC$ (1)

Where;

A = design cement content, percent total weight of soil

B = percent passing 400-micron sieve size, expressed as a decimal

C = percent cement required to obtain the desired PI of minus 400-micron material, expressed as a decimal.

The second mechanism has been identified as adhesion. The aggregate particles are held to the surface by the asphalt, which acts as a binder or cement. To increase shear strength, the cementing effect increases cohesiveness. Stability and gradation requirements form nearly all of the design criteria for bituminous stabilized soils and aggregates. The freeze-thaw and wet-dry durability tests are meaningless for mixtures stabilized with asphalt. Bituminous stabilization is a better option for hot, dry areas. Bituminous binder is added to the soil to reduce water penetration and provide cohesiveness to non-plastic materials. The best options for bituminous stabilization are granular and easily granulated materials. When bitumen-stabilized material is used as the pavement's wearing course, there are limitations. Unless substantial amounts are applied, the binding action of bitumen alone will not be sufficient to stop traffic and weathering while traveling. Such a high bitumen content is usually not cost-effective. Successful applications of bituminous stabilization have been seen in the handling of crushed rock, gravel, sandy loam, sand-clays, and other materials. With increasing amounts of material passing through a 75-micron filter, bitumen can be used to stabilize fine-grained soils; however, doing so will increase costs and the quantity of asphalt material needed. Materials with a plasticity index of less than 10% make the best candidates for this kind of stabilization (Al Tabbaa and Stegemann, 2005; Kowalski and Starry, 2007).

1.4.6. Stabilization with Lime-Fly Ash (LF) and Lime-Cement-Fly Ash (LCF)

Coarse-grained soils with little to no fines can often be stabilized by using LF or LCF combinations. Fly ash, sometimes referred to as coal ash, is a mineral byproduct that is created when pulverized coal is burned. When mixed with lime and water, the compounds of silicon and aluminum present in it solidify into a high-compressive-strength cement mass. Fly ash acts as an agent that the lime can react with, so the two are commonly used together with success to stabilize granular materials. Any type of sand, gravel, or combination of sand and gravel soil can be stabilized using lime-fly ash or lime-cement-fly ash combinations. These soils should have a Plasticity Index of no more than 25% and no more than 12% fines. It is therefore often acceptable to use LF or LCF for stabilization when using base and subbase course materials. The stabilization process with cement or lime differs slightly from the LF design. During the mix design process, the percentage of lime-fly ash, the moisture content, and the ratio of lime to fly ash can all be adjusted for a particular material combination (aggregate, fly ash, and lime). It is commonly acknowledged that engineering attributes like strength and durability are directly impacted by the matrix material's quality. The matrix material is the portion composed of fly ash, lime, and fine aggregate particles. Basically, stronger and more durable results are possible when the matrix material can "float" the coarse aggregate particles.

The fine-sized particles efficiently fill in the gaps left by the coarse aggregate particles. For each coarse aggregate material, a specific amount of matrix is required to effectively fill the available void spaces and enable the coarse aggregate particles to "float." The amount of matrix required to reach the combination's MDD is known as the optimum fines content. In LF combinations, it is recommended that the amount of matrix be approximately 2% higher than the optimal fines content. The strength development at the concentration of the permissible fine is also influenced by the lime to fly ash ratio. By altering the ratio of lime to fly ash, different values for strength and durability can be obtained.

1.4.7. Stabilization with Ground Granulated Blast Furnace Slag (GGBS)

A byproduct of the blast furnaces used to produce iron is ground granulated blast furnace slag or GGBS. They are fed with an exact mixture of limestone, coke, and iron ore and operate at a temperature of about 1500°C. After the iron ore is transformed into iron, the remaining ingredients form a slag that floats on top of the iron. If used in the creation of GGBS, this slag is frequently tapped off as a molten liquid and needs to be rapidly cooled in a lot of water. Quenching maximizes the properties of the cementation and produces granules that resemble coarse sand. This "granulated" slag is then dried and ground to a fine powder to a precisely controlled fineness in sophisticated manufacturing facilities capable of processing up to 500,000 tons of slag annually. Even though GGBS powder sets relatively slowly by itself, alkali is typically needed to activate and speed up the cement's setting time for practical use. Since GGBS alone only has slow cementation properties, Portland cement frequently supplies the alkalinity to activate and accelerate these capabilities. Another way to get the alkali needed for activation is to use lime. Laboratory and field studies have shown that both sulfides and sulfates can cause a disruptive expansion in stabilized soils. It has been shown that GGBS and lime combinations are effective and practical for stabilizing soil while also providing technical benefits. In particular, the integration of GGBS works very well to stop the expansion caused by sulfate or sulfide in the soil (Al Tabbaa and Stegemann, 2005; Kowalski and Starry, 2007).

1.4.8. Stabilization with Geotextiles

Geotextiles can help spread loads and thereby increase the site's load-bearing capacity over very soft soils because of their tensile strength capabilities. It is necessary to use a geotextile as a separation layer before adding any cover aggregate to a soil sample that has more than 10% fine particles. Along with serving as a water barrier to keep moisture out of the pavement, geotextiles can also function as separators to stop excess fines from passing through granular material that has been placed over them. By using geotextiles as filter media to create various drainage layers within and next to the pavement, excess moisture can be managed and removed. Especially for expedient applications, the use of geotextiles will facilitate traffic movement over low-bearing capacity soils. Geotextiles have the potential to reduce or even completely eliminate the need for more conventional stabilization materials. When using a geotextile for separation, it should fulfill the drainage or filtration specifications for the particular soil circumstances. Appropriate sizing of the geotextile openings is necessary to prevent soil particle migration. In subgrade conditions and when covering arterials, the geotextile must be strong enough to meet survivability requirements. When using geotextile as a water barrier or reinforcement material, there are design guidelines that must be followed. To act as a water barrier, the geotextile frequently needs to be coated with bitumen. Although there are several ways to field sew geotextile sheet seams together, to prevent problems with fastening, they are usually only partially overlapped.

1.4.9. Stabilization Using Fibers and Prefabricated Materials

Utilizing a pulverize mixer, hair-like fibers are mixed into the moist soil to stabilize it. The best soil types for fiber stabilization include sands and salty sands that fall into the categories of SW, SP, SM, and some SM-SC soil types. Fibers should normally only be applied to the previously mentioned coarse-grained soil types because their use in high-plasticity soils has led to inconsistent results. For soil stabilization, examples of manufactured materials that can be used as a trafficked surface to support loads on soft soil include Hex-Mats, Uni-Mats, and any other type of material.

1.4.10. Stabilization with Rice Husk Ash and Lime Sludge

As byproducts, a variety of industries worldwide produce large amounts of waste, such as lime sludge and rice husk ash. These pollutants pose a significant disposal challenge and have detrimental effects on the surrounding areas and environment. By utilizing this waste material in the construction of roads, the problem of their disposal can be significantly diminished. Numerous researchers have studied the application of rice husk ash in stabilizing soil masses, and their findings indicate that this technique significantly improves the properties of the soil. Certain studies indicate that it is especially beneficial for stabilizing clayey soils.

The results of some studies are given below:

- It increases the liquid limit and plastic limit thereby decreasing the PI value of the soil
- It increases the unconfined compressive strength of soil.
- It increases the soaked CBR of the soil.
- The optimum proportioning of lime sludge and rice husk ash for maximum unconfined compressive strength and lowest plasticity index is 16% and 10% respectively.
- The soaked CBR however kept on increasing at 15% and 20% rice husk ash (Al Tabbaa and Stegemann, 2005; Kowalski and Starry, 2007).

REFERENCES

1. Addo, J. Q., Sanders, T. G., & Chenard, M. (2004). Road dust suppression: Effect on unpaved road stabilization.
2. Al Tabbaa, A., & Stegemann, J. A. (2005). *Stabilization/Solidification Treatment and Remediation: Proceedings of the International Conference on Stabilization/Solidification Treatment and Remediation*. CRC Press.
3. Baghini, M. S., et al. (2017). Effects on engineering properties of cement-treated road base with slow-setting bitumen emulsion. *International Journal of Pavement Engineering, 18*(3), 202–215.
4. Cizer, Ö., et al. (2006). Carbonation reaction of lime hydrate and hydraulic binders at 20°C. In *Proceedings of the First International Conference on Accelerated Carbonation for Environmental and Materials Engineering*. The Royal Society.
5. Cortellazzo, G., & Cola, S. (1999). Geotechnical characteristics of two Italian peats stabilized with binders. In *Proceedings of Dry Mix Methods for Deep Soil Stabilization* (pp. 93–100).
6. Das, B. M. (2003). Chemical and mechanical stabilization. *Transportation Research Board*. ANSI.
7. EuroSoilStab. (2002). *Development of Design and Construction Methods to Stabilize Soft Organic Soils: Design Guide for Soft Soil Stabilization*. CT97-0351, European Commission, Industrial and Materials Technologies Programme (RITE-EURAM III).
8. Grogan, W. P., Weiss, C. A., Jr., & Rollings, R. S. (1999). *Stabilized Base Courses for Advanced Pavement Design: Report 1—Literature Review and Field Performance Data*. Army Engineer Waterways Experiment Station, Vicksburg, MS: Geotechnical Laboratory.
9. Hall, M. R., Najim, K. B., & Keikhaei Dehdezi, P. (2012). Soil stabilization and earth construction: Materials, properties and techniques. In *Modern Earth Building* (pp. 222–255).
10. Hebib, S., & Farrell, E. R. (1999). Some experiences of stabilizing Irish organic soils. In *Proceedings of Dry Mix Methods for Deep Soil Stabilization* (pp. 81–84).
11. Jawad, I. T., et al. (2014). Soil stabilization using lime: Advantages, disadvantages, and proposing a potential alternative. *Research Journal of Applied Sciences, Engineering and Technology, 8*(4), 510–520.
12. Kearney, E., & Huffman, J. (1999). Full-depth reclamation process. *Transportation Research Record: Journal of the Transportation Research Board, 1684*, 203–209.

13. Keller, I. (2011). Improvement of weak soils by the deep soil mixing method. *Keller Brochure* (pp. 23–30).
14. Kowalski, T. E., & Starry, D. W., Jr. (2007). Modern soil stabilization techniques. In *Characterization and Improvement of Soil Materials Session*, Annual Conference of the Transportation Association of Canada, Saskatoon, Saskatchewan, Canada.

CHAPTER 2

SOIL CLASSIFICATION AND TESTING FOR STABILIZATION

Contents

INTRODUCTION

Testing and classifying the soil are important steps in the stabilization process, which is necessary to guarantee the stability and longevity of building projects. Sorting soil into different groups according to its chemical and physical characteristics is known as soil classification. The Unified Soil Classification System (USCS) and the American Association of State Highway and Transportation Officials (AASHTO) classification system are the two main classification schemes. The USCS classifies soil into further subcategories according to grain size and plasticity, including gravel, sand, silt, and clay. The AASHTO system uses particle size and engineering properties to classify soil, mainly for use in highway construction. Accurate classification aids in determining the behavior of the soil under loading circumstances for creating stable pavements and foundation.

Equally important to soil stabilization is testing, which offers empirical data to evaluate soil properties and direct the choice of suitable stabilization techniques. Common tests include the compaction test, which establishes the maximum dry density (MDD) and ideal moisture content for reaching the intended soil compaction, and the Atterberg limits test, which gauges the soil's plasticity. The strength and load-bearing capacity of the soil are assessed using the California Bearing Ratio (CBR) test, which is essential for designing roads and pavements. Furthermore, the soil's resistance to shear forces is evaluated by the shear strength test, which includes direct shear and triaxial tests. This gives information about the soil's stability under load.

The techniques for stabilizing soil differ based on the kind of soil and the needs of the project. Blending soils to increase their shear strength and compaction is known as mechanical stabilization. By causing chemical reactions that bind soil particles and strengthen them, additives like lime, cement, or fly ash are used in chemical stabilization to improve the properties of the soil. Plant roots and organic matter are incorporated into biological stabilization to enhance soil stability and structure. For each method, exact soil testing is necessary to ascertain the right mix and rates of application.

A key component of the soil stabilization process is the testing and classification of the soil. Through precise soil classification and rigorous testing, engineers can choose the best stabilization strategies to improve soil performance and guarantee the longevity of construction projects. When carried out correctly, these procedures reduce the dangers brought on by unstable soil and enhance the general security and performance of engineering projects.

2.1. OVERVIEW OF SOIL CLASSIFICATION SYSTEMS

Road construction relies heavily on soil classification systems because they offer a methodical framework for assessing soil qualities and figuring out whether they are suitable for supporting road infrastructure. The Unified Soil Classification System (USCS) and the American Association of State Highway and Transportation Officials (AASHTO) system are the two most widely used classification schemes in the road construction industry. Each system provides distinct insights into the behavior of soil, assisting engineers in creating sturdy and stable roads.

Soils are categorized by the USCS according to their characteristics of plasticity and particle size. It separates soils into four major categories: clay, silt, gravel, and sand. Texture and plasticity are used to further categorize each of these primary groupings. For example, there are two categories for gravel: coarse and fine, and there are three categories for sand: coarse, medium, and fine. Clays and other higher plasticity soils are classified according to their Atterberg limits, which gauge the soil's capacity to flex under pressure. Understanding how various soils will behave under load—a crucial skill for ensuring that the subgrade and embankments built for roads can support traffic loads without experiencing excessive deformation or failure—is best achieved with the help of the USCS.

The AASHTO system, in comparison, places more emphasis on a soil's suitability for road construction, especially about its drainage and load-bearing capabilities. It divides soils into groups called A-1 through A-7, where A-1 soils are the best quality and A-7 soils are the least suitable for building roads. Particle size distribution, plasticity, and compaction characteristics are the three main criteria used to classify these materials. For instance, A-1 soils are perfect for use as a subbase or base layer in road construction because they are usually well-graded and have low plasticity. Conversely, A-7 soils, which frequently have a high clay content, might need to be stabilized or treated to improve their engineering qualities.

Tests are used in both systems to give a comprehensive understanding of soil properties. To ascertain particle size distribution and plasticity for the USCS, procedures such as the sieve analysis, hydrometer analysis, and Atterberg limits tests are employed. Although other tests like the Proctor compaction test and the California Bearing Ratio (CBR) test are used to evaluate soil strength and compaction, respectively, sieve analysis is still an important part of the AASHTO system. The outcomes of these tests help determine what needs to be treated or stabilized to improve soil performance and guarantee that the soil satisfies the requirements for building roads.

The systems for classifying soils are vital for road construction because they offer a thorough understanding of the characteristics and behaviors of the soil. While the AASHTO system evaluates soil suitability based on load-bearing capacity and compaction characteristics, the USCS provides insights into soil texture and plasticity. Engineers can ensure the construction of stable, dependable, and long-lasting roads by using these classification systems and related tests to help them make educated decisions about stabilization and soil treatment (Al-Tabbaa and Evans, 2005; EuroSoilStab, 2002; Porbaha et al., 2005).

2.1.1. AASHTO Soil Classification System

The American Association of State Highway and Transportation Officials created the AASHTO Soil Classification System, which serves as a manual for classifying soils and soil-aggregate mixtures for use in highway construction.

The classification system was first developed by Hogentogler and Terzaghi in 1929 but has been revised several times since (Table 2.1).

Table 2.1. *AASHTO Soil Classification System (from AASHTO M 145 or ASTM D3282)*

General Classification	**Granular Materials (35% or Less Passing the 0.075 mm (No. 200) Sieve)**							**Silt-Clay Materials (>35% Passing the 0.075 mm (No. 200) Sieve)**			
Group Classification	A-1		A-3	A-2				A-4	A-5	A-6	A-7
	A-1-a	A-1-b		A-2-4	A-2-5	A-2-6	A-2-7				A-7-5 A-7-6
Sieve Analysis, % passing											
2.00 mm (No. 10)	50 max	–	–	–	–	–	–	–	–	–	–
0.4255 mm (No. 40)	30 max	50 max	51 min	–	–	–	–	–	–	–	–
0.0755 mm (No. 200)	15 max	25 max	10 max	35 max	35 max	35 max	35 max	36 min	36 min	36 min	36 min
Characteristics of fraction passing 0.425 mm (No. 40)											

Liquid Limit	–	–	40 max	41 min	40 max	41 min	40 max	41 min	40 max	41 min
Plasticity index	6 max	N.P.	10 max	10 max	11 min	11 min	10 max	10 max	11 min	11 min[1]
Usual types of significant constituent materials	stone fragments, gravel and sand	fine sand	silty or clayey gravel and sand				silty soils		clayey soils	
General rating as a subgrade	excellent to good						fair to poor			

The plasticity index of the A-7-5 subgroup is equal to or less than the LL – 30. The plasticity index of the A-7-6 subgroup is greater than LL – 30.

Source: https://en.wikipedia.org/wiki/AASHTO_Soil_Classification_System.

2.1.2. Unified Soil Classification

An essential tool in road building is the USCS, which provides a methodical way to classify and evaluate soil according to its engineering and physical characteristics. The USCS was created in the 1950s and is frequently used to evaluate the behavior of soil and its suitability for different construction applications, such as road infrastructure.

Based on factors like grain distribution, plasticity, and particle size, the USCS divides soil into many main categories. The system classifies soils into four main groups: clay, silt, gravel, and sand. More divisions within each category offer more precise classifications. For example, sand is classified as coarse, medium, and fine, and gravel is divided into coarse and fine. Tests like hydrometer analysis, which evaluates the fine particle content, and sieve analysis, which gauges the distribution of particle sizes, are used to establish these classifications.

The term "plasticity" describes the soil's capacity to change without breaking or becoming weaker. The liquid limit (LL) and plastic limit (PL) are two of the Atterberg limits that are used to quantify this. The plasticity index (PI) is defined as the variation between these limits. High plasticity soils, like clays, can experience large volume changes in response to variations in moisture content, which can negatively impact the performance and stability of roads. On the other hand, under varied moisture conditions, soils with low plasticity—such as granular materials—generally show more stable behavior.

A system of soil symbols and descriptive terms is also included in the USCS to aid in the documentation and communication of soil properties. For instance, the symbols CL stand for low-plasticity clay and GW for well-graded gravel. Engineers can quickly identify and evaluate soil types and their suitability for various construction applications with the aid of these symbols.

Several important design and construction decisions in road construction are influenced by precise soil classification based on the USCS. For subbase and base layers, for example, well-graded gravel or sand (GW or SW) is frequently chosen because of its favorable compaction and drainage qualities. On the other hand, clays (CL or CH), especially those with high plasticity, might need to be stabilized or treated to increase their ability to support loads and lessen the likelihood that they will expand or contract.

Data from USCS-related testing techniques, like sieve analysis and Atterberg limits, are crucial for classifying soils. These tests aid in identifying the proper techniques for stabilizing or compacting the soil to improve soil performance and guarantee the stability of the road infrastructure.

The USCS provides a thorough and organized method for comprehending soil properties, making it a vital tool in the road construction industry. The USCS assists engineers in evaluating soil behavior, directing material selection, and making well-informed decisions to guarantee the stability, longevity, and functionality of roads by categorizing soils according to particle size and plasticity (Al-Tabbaa and Evans, 2005; EuroSoilStab, 2002; Porbaha et al., 2005).

2.1.3. Principle of Soil Classification

Any classification serves the function of organizing our knowledge so that, for the sake of a particular goal, the characteristics of objects may be most easily remembered and their relationships understood. Classes are created during the process by putting objects together based on shared characteristics. Groups that allow for the most accurate, significant, and numerous statements to be made for the goal serve the purpose most effectively in any classification system. Classification aids in managing complexity. There are too many items to take into account separately. We can create useful classes to assist us in organizing our knowledge and streamlining our decision-making if we can identify any shared characteristics or behaviors between them.

- What is Different About Soils?
  - There is really no soil 'individual' as a self-standing object
  - The concept of the Pedon as a discrete object within the soil continuum
  - Therefore, the emphasis is on defining capable classes rather than on the optimal classification of individuals.
  - There is no true inheritance or genetics as it is understood in biology
- Major Ways of Classifying Soils

There are various ways to organize a soil classification. A major distinction is between natural and technical approaches:

- Natural soil classifications group soils by some intrinsic property, behavior, or genesis of the soils themselves, without reference to the use
- Technical soil classifications group soils by some properties or functions that relate directly to a proposed use or group of uses.

- Natural Classifications
  - Group by ecologic region, e.g., "prairie soils," or "boreal soils."

Geographically-compact but may have diverse properties and functions.

- Group by presumed genesis, i.e., the development pathway of the soil profile. These are called genetic soil classifications. The soil individual is considered a natural body with its own history and ecology. This depends on the interpretation of landscape and soil genesis.
- Group by similar properties, working bottom-up from a set of individuals to a set of classes, and then grouping the classes into super-classes. This can be done by subjective judgement of the classifier; nonnumerical classification, usually multivariate.

- Technical Classifications
  - Hydrologic response
  - Suitability classes (FAO Framework for Land Evaluation)
  - Land Use Capability (USDA LCC)
  - Fertility Capability Classification (FCC)
  - Engineering group

If the groups that emerged in this manner matched, that would be really lovely. For instance, it stands to reason that soils with comparable genesis would exhibit comparable characteristics and behaviors. The early classifiers of soil believed that the origins and characteristics of all soils in an ecological region were the same. But it's not always the case.

2.1.4. International Soil Classification Systems

These are systems designed for universal application. They should classify any soil, and serve to correlate experiences on similar soils all over the world.

2.1.4.1. World Reference Base for Soil Classification Background

The International Union of Soil Sciences has approved this system as the global standard for classifying soil. It was created through an international partnership supported by the FAO through its Land and Water Development division, the International Union of Soil Science (IUSS), and the International Soil Reference and Information Centre (ISRIC). The WRB, which takes its cues from contemporary soil classification theories such as soil taxonomy, the FAO Soil Map of the World 1988 legend, the Référentiel Pédologique, and Russian notions, has replaced the FAO Legend for the Soil Map of the World. It is therefore the outcome of "pedo-political" discussions. The primary criterion for classification is soil morphology, which is believed to convey the impacts of soil genesis. Soil climate is not included in the Soil Taxonomy system, except how soil properties are impacted by climate. This is one of the main differences between the two systems. To ensure the simplest possible correlation with national and international systems, diagnostic criteria are aligned as closely as possible with those of the current systems.

- Classification Principles:
  - A practical grouping into reference groups that share an assemblage of features that cause distinct behavior, i.e., ecological function and implications for soil management.
  - So, the reference groups are quite distinct and can serve as the basis for making general statements about soil suitability and management strategies.
  - Second-level qualifiers are also based on practical considerations, allowing us to make more specific statements about soil behavior.
  - Soil Climate is not taken into account, except insofar as it affects observable soil properties.
- Structure: The WRB is a Two-Level Classification:

1. **Reference Soil Groups:** Intermediate in conceptual level between Soil Taxonomy orders and suborders; Examples: Histols, Fluvisols, Luvisols
2. **Second-Level Subdivisions:** Using any defined combination of 121 qualifiers. It is possible to use either a single qualifier (the most important) or all relevant qualifiers, depending on the degree of detail needed. Similar in detail to Soil Taxonomy great groups (one qualifier) or subgroups (multiple qualifiers).

- Examples: Leptic Umbrisols, Chromi-Vertic Luvisols:

The subdivisions do not account for every potential variation between soil map units. Specifically, the following factors are not taken into consideration as such: slope, geomorphology, erosion, parent material, vegetation, depth of water table, drainage, and physiographic features, unless they have an impact on soil morphology. Although these characteristics can be used locally to specify mapping phases, they are not thought to be suitable for classification as soil properties. At this level of detail, certain internal characteristics are also ignored, such as the thickness and morphology of the solum, individual horizons, and substratum layers. For an in-depth soil survey, these can be used to define series or forms locally.

For detailed mapping and site characterization, local organizations or survey projects are expected to use locally-defined soil series, soil forms, or similar. The WRB is used to group these locally-defined soils for correlation and communication. So, a soil survey using the WRB would also need a local classification. A typical structure is:

WRB ⇒ Soil Forms or Families ⇒ Soil Series ⇒ Mapping phases

The families may well take into account soil or atmospheric climate, according to the aims of the survey.

- Classifying with WRB

1. **Assemblages of Characteristics:** which diagnostic horizons are present? (both surface and subsurface)

E.g., 'mollic horizon,' 'cambic horizon.'

2. **Assemblages of Characteristics:** diagnostic soil properties and materials as required by the keys

E.g., 'alic properties,' 'abrupt textural change'; 'fluvic soil material.'

3. **Reference Soil Groups:** single-level binary-elimination key.

Structure of a binary-elimination key ('other soils having...'); reading a key (pay attention to and or)

4. Second-level subdivisions:
   - List of possible second-level names for each reference group.
   - Description of criteria for each second-level name.
   - Can use one (most important) or all applicable.
5. Procedure to follow
   - Pre-conception OK, to get some idea of where we might end up and to avoid blunders, however.
   - Must follow the exact wording of the keys to classify.
   - Determine diagnostic surface and subsurface horizons and their thicknesses.
   - Follow the key to the reference groups; during this, you may be asked to determine if the soil satisfies certain diagnostic criteria such as the presence of soil properties or materials.
   - For each possible second-level name of the reference group, see if the soil meets the criteria. List them in priority order.

2.1.5. The USDA System for Soil Classification

Soil taxonomic classifications reflect the dominant soil forming factors active during soil formation at a particular location. The USDA system of Soil Taxonomy (soil naming) consists of a hierarchy of six levels. These levels, in order from most general to most specific, are:

- Order
- Suborder
- Great Group
- Subgroup
- Family
- Series

2.1.5.1. Soil Taxonomy

It is similar to the kingdom, phylum, class, order, family, genus, and species classification scheme used in biology (Linnean system). Land managers can use the information provided by this system of soil classification to conclude the suitability of a specific soil for plant production, urban and residential uses, waste management, and construction sites.

2.1.5.1. Taxonomic Classification of the Soils

The system of soil classification used by the National Cooperative Soil Survey has six categories. Beginning with the broadest, these categories are the Order, Suborder, Great Group, Subgroup, family, and Series.

These categories are defined in the following paragraphs:

- **Order** – 12 soil orders are recognized. The differences among orders reflect the dominant soil-forming processes and the degree of soil formation. Each order is identified by a word ending in 'sol.' An example is Alfisols.
- **Suborder** – Each order is divided into suborders primarily based on properties that influence soil formation and/or are important to plant growth.
- **Great Group** – Each suborder is divided into great groups on the basis of similarities in horizons present, soil moisture or temperature regimes, or other significant soil properties.
- **Subgroup** – Each great group has a 'typic' (typical) subgroup which is basically defined by the Great Group. Other subgroups are transitions to other orders, suborders, or great groups due to properties that distinguish them from the great group.
- **Family** – Families are established within a subgroup based on

physical and chemical properties along with other characteristics that affect management.

- **Series** – The series consists of soils within a family that have horizons similar in color, texture, structure, reaction, consistency, mineral and chemical composition, and arrangement in the profile (Al-Tabbaa and Evans, 2005; EuroSoilStab, 2002; Porbaha et al., 2005).

2.2. SOIL TESTING METHODS FOR ROAD CONSTRUCTION

The design, construction, and longevity of road infrastructure are all directly impacted by the properties and behavior of the soil, which is why soil testing techniques are essential to the process of building new roads. Precise soil testing guarantees that the subgrade and structural layers of the road possess the strength, stability, and drainage qualities required to sustain traffic volumes and endure weather conditions.

Key soil testing methods commonly employed in road construction:

- **Sieve Analysis:** The process of sieve analysis is essential for ascertaining the soil's particle size distribution. A soil sample is put through a succession of sieves with progressively smaller apertures as part of this test. Engineers can create a particle size distribution curve by weighing the soil that is retained on each sieve. This allows them to categorize the soil based on its texture (e.g., clay, silt, gravel, or sand). Understanding how the soil will behave under loading and compaction circumstances is essential, and this information informs choices about material selection and treatment.
- **Atterberg Limits:** The fine-grained soils' (like clays') plasticity is measured by the Atterberg limits test. The LL, PL, and PI are its three main parameters. The water content at which soil transitions from a liquid to a plastic state is known as the liquid limit, and the water content at which soil transitions from a plastic to a semi-solid state is known as the plastic limit. The difference between the liquid and plastic limits is known as the plasticity index. These limitations shed light on how the soil behaves in different moisture scenarios and whether it is suitable for use in road building.
- **Proctor Compaction Test:** The ideal moisture content at which a soil achieves its MDD when compacted is found using the Proctor compaction test. This test is essential to make sure aggregate or soil used in road construction is compacted to a density that offers sufficient stability and strength. Using a standard energy level, soil

samples are compacted into a mold for the test, and the resulting dry density is then measured. The findings aid in figuring out the right moisture content to achieve the required level of compaction in the field.

- **California Bearing Ratio (CBR) Test:** Designing road bases and subgrades requires careful consideration of the strength and load-bearing capacity of the soil, which is assessed by the CBR test. This test involves compacting a sample of soil to a predetermined density and then allowing a cylindrical plunger to pierce it. The measurement and comparison of the resistance to penetration is done against a standard reference material, usually crushed rock. The CBR value guides the design of road pavement thickness and material requirements by indicating the soil's capacity to support traffic loads.
- **Shear Strength Tests:** The resistance of the soil to shear forces is measured by shear strength tests, such as the direct shear test and the triaxial shear test. Applying a horizontal force to a soil sample in a shear box until failure happens is how the direct shear test determines the shear strength of the soil. To find the shear strength of a soil sample under more complicated loading conditions, the triaxial shear test entails applying axial and confining pressures to the sample. The sample should be cylindrical. These tests yield useful information for designing embankments and slopes as well as assessing soil stability.
- **Consolidation Test:** The compressibility and rate of settlement of the soil under applied loads are measured by the consolidation test. A soil sample is put inside a consolidation device, where it is gradually subjected to vertical loads while the sample's height is being tracked. The findings aid in estimating the amount of settlement that will transpire over time, a crucial aspect of designing sturdy road foundations and steering clear of excessive settlement that might give rise to structural problems.
- **Permeability Test:** The permeability test evaluates the soil's capacity to transfer water, an essential component in guaranteeing appropriate drainage when building new roads. To help engineers design drainage systems that prevent water accumulation and lower the risk of road damage from water-induced erosion or weakening, the test measures the rate at which water flows through a soil sample.

To assess the technical and physical characteristics of soils utilized in road building, soil testing techniques are crucial. Through the provision of vital information on soil classification, compaction, strength, and drainage, these tests help engineers make well-informed decisions and guarantee the building of sturdy, stable, and dependable roads.

2.2.1. Important Soil Tests for Construction

Examining the physical, mechanical, and chemical characteristics of the soil before starting construction is known as soil testing. To guarantee the quality of your building structures, soil testing is crucial during the construction process. It provides a quick overview of the soil's bearing capacity. To ascertain the various soil parameters, a variety of soil testing techniques are used.

In the construction industry, soil testing is very important, so all contractors must understand this vital process. To guarantee the structural integrity, stability, and safety of any building or infrastructure project, it is essential to comprehend the qualities and features of the soil at the construction site.

Contractors gain the most from determining the type of soil that is present on the site. When exposed to external forces, different soil types—such as clay, silt, sand, or their combinations—display differing load-bearing capacities and behaviors. Contractors can ensure the structure can withstand expected loads and environmental conditions by making informed decisions regarding foundation requirements and design with the knowledge of the type of soil (Figure 2.1).

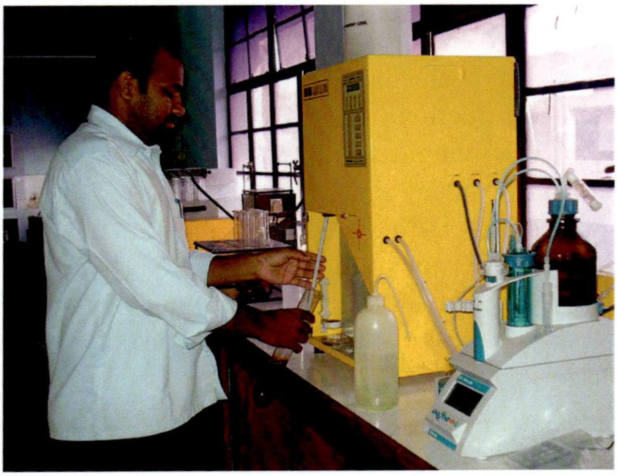

Figure 2.1. *Soil tests.*

Source: By Balaji Kasirajan – Own work, CC BY-SA 3.0, https://commons.wikimedia.org/w/index.php?curid=28916350.

Furthermore, soil testing offers important information about the moisture content and compaction of the soil. In order to avoid soil instability, differential settlement, or settling, which can result in structural damage and safety risks, adequate compaction is essential. Through soil testing, contractors can ascertain the ideal degree of moisture and compaction needed for a stable foundation, allowing them to modify their building techniques accordingly.

Moreover, soil testing enables contractors to recognize possible problems like contaminated soil, expansive soils, or high-water tables. Due to moisture fluctuations, expansive soils can cause significant foundation movement, which can lead to cracks and structural damage. Elevated water levels can hinder appropriate drainage and jeopardize the stability of structures. Contamination of the soil can affect the construction's long-term durability and present environmental risks. Contractors can reduce these risks by doing soil tests and making the necessary adjustments to the design and construction process.

Furthermore, choosing the right foundation size and depth requires an understanding of the soil's bearing capacity. Contractors can design foundations that efficiently distribute the weight of the structure with the help of soil tests, which offer useful information on the load-bearing capacity. This guarantees that the foundation can withstand the anticipated loads and guards against settlement or structural collapse (Al-Tabbaa and Evans, 2005; EuroSoilStab, 2002; Porbaha et al., 2005).

2.2.2. Popular Soil Testing Methods

To determine the soil profile for a given type of soil test, a site investigation must be conducted. The characteristics of the soil at the subgrade level are necessary for road construction projects. To build roads, common soil tests include those for classifying soil, determining particle size distribution, determining moisture content, determining specific gravity, and testing for liquid and plastic limits. Soil properties, including saturation level, are determined by testing soils for moisture content, particle size, and specific gravity. There are two types of soil tests: in-situ and laboratory. Every sample that is taken to determine the moisture content and particle size should undergo the laboratory tests (Figure 2.2).

Different soil parameters are tested before construction, such as its chemical composition, density, specific gravity, and load-bearing capacity.

The various types of soil testing in construction are as follows:

Figure 2.2. *Popular soil testing methods.*

Source: By Jugewilson – Own work, CC BY-SA 4.0, https://commons.wikimedia.org/w/index.php?curid=59446538.

2.2.2.1. Dry Density Test

It is done to determine the density of soil particles per unit volume of soil excluding the moisture content. The dry density test is carried out in two ways:

1. **Core Cutter Method:** In the core cutter method, a cylindrical cutter of specific dimensions is used. The core cutter is inserted into the soil and a portion of the soil is extracted and weighed. The water content of the soil is determined, and the density of the soil is calculated through the following formula:

$$P = (M/V)/1 + W$$

2. **Sand Replacement Method:** Using a metallic sieve, some soil is removed from the ground to determine the dry density of the material. To obtain this soil, a circular hole must be made in the ground. Sand with a known density is used to fill in the hole. The dry density of the soil is determined by computing the hole's volume.

2.2.2.2. Moisture Test

The purpose of this test is to find out how much moisture is in the soil that will be used to lay the foundation. The oven drying method, the calcium carbide method, the radiation method, the sand bath method, and the alcohol method are some of the techniques used for moisture testing.

Numerous techniques, including the alcohol method, radiation method, sand bath method, Pycnometer method, torsion balance method, calcium carbide method, and oven-drying method, can be used to measure the moisture content of the soil. The oven-drying method, which involves weighing the sample soil, drying it at 110°, and then weighing it again to determine the difference, is the most widely used technique for determining moisture content.

To find out how much water is in the soil, various moisture content tests are frequently used in the construction industry. These tests support appropriate compaction, stable foundations, and high-quality construction materials overall.

Here are some of the key types of moisture content tests for construction:

- **Gravimetric Method (Oven-Drying Method):** This is one of the most accurate and widely used methods. It involves collecting a soil sample, weighing it when wet, drying it in an oven at a specific temperature (typically around 110°C), and then reweighing it to calculate moisture content based on the weight difference.
- **Calcium Carbide Method:** In this method, soil samples are mixed with calcium carbide, which reacts with water to produce acetylene gas. The amount of gas produced is proportional to the moisture content in the soil.
- **Torsion Balance Method:** The torsion balance method measures the weight loss of a soil sample during drying. It is suitable for both field and laboratory applications.
- **Pycnometer Method:** This method involves using a calibrated glass container (pycnometer) to measure soil moisture content. The sample is saturated with water, and the change in volume is used to calculate moisture content. It is often used in laboratory settings due to its precision.

The moisture content test provides critical information about the soil's water content, which is essential for various engineering calculations and the design of foundations, slopes, and retaining structures.

2.2.2.3. Atterberg Limits Test

To ascertain the soil's liquid limit, plastic limit, and plasticity index, the Atterberg limits test is used. The purpose of the liquid limit test is to ascertain the soil's consistency and moisture content at which it behaves more like a viscous liquid than a plastic clay. The soil liquid limit is measured with a Casagrande device.

The point at which the soil behaves like a brittle solid is determined by its plastic limit.

This method of testing soil involves spreading the soil on a glass sheet after it has been combined with water. A hand-rolled roll of soil is formed slowly until it breaks and crumbles. To find the soil's plastic limit, the water content of the soil is measured at the point at which the crumbles at a diameter of 3 mm. The plastic limit is subtracted from the liquid limit to determine the soil's plasticity index.

2.2.2.4. Specific Gravity Test

The ratio of soil density to water density is used to determine the specific gravity of the soil. There are several methods for figuring out the specific gravity, including the shrinkage limit method, gas jar method, pycnometer method, and density bottle method. This process of testing the soil establishes how vulnerable it is to corrosion and how likely it is to float away during floods.

2.2.2.5. Soil Tests for Construction

The purpose of this test is to determine the soil's compaction. To densify the soil, it decreases air voids. A common laboratory test used in the construction industry to ascertain the ideal moisture content and MDD of soil is the Proctor's compaction test. Concerning planning and building earthwork projects like embankments, roads, and foundations, this test helps engineers and contractors assess the soil's compaction and compatibility.

A uniform protocol for the Proctor's compaction test is described in numerous national and international standards, including ASTM and IS codes. A representative soil sample is compacted under controlled moisture content and compaction effort conditions in a compaction mold as part of the test.

- **Step 1:** The first step of the test involves obtaining a representative soil sample from the site. The sample is then dried, crushed, and sieved to eliminate coarse particles or organic materials. The resulting soil sample is divided into different moisture content increments.
- **Step 2:** Next, the compaction mold, typically in a cylindrical shape, is filled with the moist soil sample in layers. Each layer undergoes compaction using a specified number of blows from a compaction hammer or a mechanical compactor. The compaction effort typically relies on the weight of the compaction hammer and the fall height. This process is repeated for each moisture content increment.
- **Step 3:** Following compaction, the compacted soil specimen is

carefully extracted from the mold and weighed to determine its mass. The moisture content of the specimen is determined by drying it in an oven and comparing the initial and final weights. The dry density of the specimen is calculated by dividing its mass by the volume of the mold.

The experiment is carried out in a range of moisture content increments, from dry to wet. Plotting the moisture content against dry density data results in a compaction curve that is also referred to as the Proctor curve. This curve can be used to determine the MDD and corresponding ideal moisture content.

Important details for the planning and execution of earthwork projects are provided by the Proctor's compaction test results. The highest density that can be attained for a particular soil under particular compaction circumstances is represented by the MDD. The moisture content at which soil can be compacted to the highest density is known as the optimum moisture content (OMC).

The Proctor's compaction test results are used by engineers and contractors to calculate the amount of compaction effort and moisture control required during construction. This helps to maximize the compaction process, ensuring long-lasting and stable structures by reaching the target density and lowering the possibility of soil-related problems or settlement.

2.2.2.6. California Bearing Ratio Test

The CBR is used to determine the soil's strength. Basically, it's done to determine whether the soil is suitable for building roads and pavement. It assists in giving information about the soil's strength. This test's methodology involves pressing a plunger into a sample of soil and then comparing the sample's resistance to a standard value.

2.2.2.7. Permeability Test

Permeability Test is the test to measure the rate at which water can flow through the soil. It helps in the designing of drainage systems and predicting the risk of soil erosion.

2.2.2.8. Soil pH Test

The purpose of this test is to determine the soil's pH level, or how acidic or alkaline it is. It guarantees the soil's chemical stability and aids in determining the likelihood that subterranean structures may corrode. Those working in environmental science, construction, or agriculture need to be well-versed in the various techniques for testing soil. It is crucial to remember that every technique has a distinct function and should be combined with other approaches for an all-encompassing evaluation. We can improve building projects, maximize crop productivity, and protect the environment for coming generations by using these soil testing techniques (Al-Tabbaa and Evans, 2005; EuroSoilStab, 2002; Porbaha et al., 2005).

2.3. DETERMINING THE SUITABILITY OF SOIL FOR STABILIZATION

A vital component of guaranteeing the longevity and structural integrity of different engineering projects, including roads, foundations, and embankments, is figuring out whether the soil is suitable for stabilization. The goal of soil stabilization is to improve the physical characteristics of soil to increase its capacity to support loads and lessen its sensitivity to external influences. Several important factors are taken into account when evaluating soil for stabilization, such as the composition of the soil, its moisture content, its compaction qualities, and the existence of contaminants or dangerous materials.

Improving the physical characteristics of soil to increase its load-bearing capacity, lessen its vulnerability to environmental influences, and guarantee the durability and safety of infrastructure projects is known as soil stabilization, and it is an essential engineering technique. Many factors, such as soil composition, moisture content, compaction characteristics, and environmental conditions, determine whether a given soil is suitable for stabilization. This thorough analysis explores these variables and outlines the procedures and techniques used to assess soil for stabilization.

2.3.1. Soil Composition and Classification

The cornerstone of soil stabilization is soil composition. A soil's particle size distribution is typically used to divide it into coarse- and fine-grained categories. Sand and gravel are examples of coarse-grained soils. They are distinguished by larger particles that usually offer good drainage but may not be cohesive. Although the stability of these soils is frequently less problematic, stabilization may be necessary to increase their ability to support loads and withstand erosion.

Fine-grained soils, like clay and silt, are more cohesive and contain smaller particles. These soils can experience large volume changes in response to variations in moisture and frequently have poor drainage, which can result in issues like shrinkage and reduced strength. To classify soils and forecast how they will behave under different circumstances, the USCS and the AASHTO classification systems are frequently utilized. Comprehending the composition and classification of soil is essential to choose the right stabilization techniques and guarantee that the soil is suitable for building.

2.3.2. Moisture Content and Plasticity

An essential component of soil stabilization is moisture content. The compaction and strength characteristics of soil are influenced by its moisture content. The amount of water needed to compact soil to its MDD is known as the optimum moisture content (OMC). Too wet soil will not reach the proper density and may become too pliable, while too dry soil will not compact well.

In the assessment of soil, the plasticity index (PI) is an additional crucial factor. The PI calculates the range of moisture content below which the soil cracks and stays plastic, or workable. Because high plasticity soils, like clays, have a propensity to expand and contract in response to changes in moisture content, stabilizing them can be difficult. Although low plasticity soils, such as silts, may not be as troublesome, compaction and stability issues can still arise. Additional treatment may be necessary to lower the plasticity and increase the stability of soil with a high PI.

2.3.3. Compaction Characteristics

Improving the strength and stability of soil requires compaction. Compaction is the process of making soil denser through the removal of air voids, improving the soil's ability to support weight. Standard Proctor Tests and Modified Proctor Tests are two common laboratory tests used to evaluate compaction characteristics.

The MDD and OMC for the soil are ascertained by these tests. A baseline for compaction specifications is provided by the Standard Proctor Test, which employs a standard compactive effort. The Modified Proctor Test is frequently applied to soils that need a higher density for optimal performance because it uses a higher compactive effort. In general, higher MDD in OMC soils makes better stabilization candidates. However, additional stabilization methods might be needed for soils that, even with correct compaction, do not reach an appropriate density.

2.3.4. Soil Stabilization Methods

The process of modifying the engineering properties of soil through chemical, biological, or mechanical means is known as soil stabilization. It is a method used in civil engineering to alter and enhance the soil's engineering qualities. Shear strength, permeability, compressibility, durability, and plasticity are some of these characteristics. Although some schools of thought prefer to use the term stabilization to refer to chemical improvements in soil qualities obtained through the use of chemical admixtures, physical or mechanical improvement is commonly used. Whether building an airport, building a road, or building a structure, the foundation of any construction project is the soil. Soil is also a crucial component of construction materials.

Once soil properties are assessed, various stabilization methods can be employed depending on the soil type and project requirements. Common soil stabilization methods include:

2.3.4.1. Mechanical Stabilization of Soil

Physical processes like compacting or tamping with equipment like rollers or rammers fall under this category. To complete this procedure, heavy machinery with the ability to apply a large compressive force is needed. This typically consists of a large compactor or, in an extreme circumstance, a crane that can lower its weight.

Often, rocky and sandy soils are stabilized by mechanical soil stabilization. Mechanical stabilization is rarely the only stabilization technique used due to developments in other forms of soil stabilization. Although mechanical soil stabilization is not commonly employed, it offers advantages. One benefit is that the soil particles undergo physical transformation rather than chemical changes, which degrade with time. The procedure is well-known for being costly, time-consuming, and difficult due to the physical alteration.

- Advantages

The advantages of mechanical stabilization of soils are:

  - Environmentally friendly.
  - Permanent.

- Disadvantages

Apart from the advantages, some disadvantages of the mechanical stabilization of soils are:-

  - Expensive.
  - Labor intensive.
  - This is no longer a popular method.

2.3.4.2. Factors Affecting the Mechanical Stabilization of Soils

The mechanical stability of mixed soil depends upon the following variables:

- **Gradation:** The gradation of the mixed soil should be such that the voids created by the coarser particles are filled with finer particles, resulting in a high density.
- **Soil Properties:** Soils with a higher liquid limit and plasticity index are generally suitable for use as soil binders for surfacing.

These soils are more cohesive and have a higher moisture retention capacity. They provide a better seal against surface water movement downward.

- **Compaction:** The mechanical stability of the stabilized soil mass is determined by the amount of compaction achieved in the field. Compaction is typically performed at or near the OMC.
- **Mechanical Strength of the Aggregates:** The mixed soil becomes more stable when the aggregates used are of high strength. However, if the mixture is properly designed and compacted, even low-strength aggregates can provide good mechanical stability.
- **Mineral Composition:** The mechanical stability of mixed soil is entirely dependent on the mineral composition. As a result, the minerals must be weather-resistant.

2.3.4.3. Chemical Stabilization of Soil

The chemical reaction that takes place between the stabilizer and the soil particles determines the chemical stabilization of the soil. Cement, lime, fly ash, bitumen emulsion, and magnesium chloride are a few of the materials used.

2.3.4.3.1. Cement or Lime Stabilization of Soil

One common technique for stabilizing soil is to use cement and lime in combination. This technique for stabilizing soil entails strengthening and resisting the soil by incorporating cement or lime into it. Depending on the characteristics of the original soil, different amounts of cement or lime are added to the soil. Generally speaking, more cement or lime is added to soil that is more plastic. Lime and cement are often combined because they are both used as binders. Even though it's a common technique, soil stabilization with cement and lime is mostly applied on paved roads. Usually, the cost of applying cement to unpaved roads is prohibitive.

Geography frequently affects how soil is stabilized with cement or lime. Because lime is easier to get in some places and not in others, cement is more affordable in those areas. The process of binding all of the soil's particles together through cement or lime stabilization strengthens the soil. Almost all soil types can be stabilized using this technique because it calls for the addition of cement or lime to the soil.

To make sure that the right amount of additives is used, soil testing is necessary. Should an insufficient quantity of additives be applied, the soil will not attain the necessary strength. Applying more could cause the soil to shrink or crack.

- Advantages

Some of the advantages of soil stabilization with cement and lime are:

  - Permanent and long-lasting.
  - It is compatible with the majority of soil types.
  - This method of soil stabilization is tested and proven.
  - Reduces the moisture content of the soil.

- Disadvantages

Apart from the advantages, some disadvantages of cement and lime stabilization of soil are:

  - Expensive.
  - Possible health risks.
  - Only applies to paved roads.
  - Comprehensive soil testing is required before application.

2.3.4.4. Soil Stabilization by Fly Ash

Although the product and method are different, fly ash soil stabilization uses a different product than lime or cement soil stabilization. Coal byproducts used in coal-fired power plants include fly ash. Although fly ash is usually kept in ponds or landfills, it can also be applied as a bonding agent to improve the soil. Similar to how cement and lime stabilization are used for soil stabilization as the sub-base of paved roads, fly ash is frequently used to treat the sub-base of paved roads. One major benefit of using fly ash regularly in soil above optimal moisture levels is that it is a dry additive.

By absorbing moisture, the powder will lower the moisture content of the soil. Fly ash stabilization of soil is usually less expensive than stabilizing soil with cement and lime. A byproduct of several different manufacturing processes is fly ash. The unique quality of fly ash as a dry additive is especially advantageous in high-moisture soils. Being a dry additive has the drawbacks of being hard to apply and potentially dangerous for workers' health because of the powder's inhalation risk. When using fly ash, there's also the curing time to take into account. The cure period is usually longer than that of soil treated with cement or lime, though it varies depending on the fly ash and soil.

- Advantages

Some advantages of soil stabilization by fly ash are as follows:

  - Long-lasting, permanent.
  - Most soil types are compatible.

- Various types for various applications.
- More affordable than cement or lime stabilization of soil.
- Useful in high moisture soils.

• Disadvantages

Some of the disadvantages of soil stabilization by fly ash are:

- Based on the availability of coal-fired power plants.
- Labor intensive method.
- Health concerns.

2.3.4.5. Polymer Stabilization

Polymer stabilization of soil is the process of adding polymers to improve the engineering and physical properties of soils. Because of their interactions with clayey soil particles, polymers often strengthen soil. Many of the polymers currently in use increase the shear strength and water retention of soil. Biopolymers and synthetic polymers are the two types of polymers used for soil stabilization. Comparing biopolymers to other chemical soil stabilizers, the environmental impact is lower.

Because polymers are based on long repeating molecules, soil particles attach themselves to the molecules. Polymers work incredibly well with any kind of soil, from clay to sandy. Because the soil particle must fit inside the polymer's molecule, the size of the polymer's molecule influences how well it performs on any given soil. The size of the polymer's molecules is particularly crucial because a molecule that is too big might prevent a small clay particle from fitting snugly, or a small molecule and a large sand particle could not fit together properly. Although polymers are a versatile class of soil stabilizers, some are not suitable for fine soils. Depending on their characteristics, polymers can have different lifespans and characteristics. Conventional treatments, on the other hand, require little maintenance and last one to three years. For the duration of the application, the majority of polymer applications produce an impermeable surface. But water performance varies as much as lifespan.

Advantages

Some advantages of polymer stabilization of soils are:

- Simple to use.
- Impermeable.
- Compatible with all types of soils.

Disadvantages

Apart from the advantages, the polymer stabilization of soil has some disadvantages. They are as follows:

- Not always good for the environment.
- Lifespan is limited.
- Each application necessitates a large number of products.

2.3.4.6. Soil Stabilization Using Geotextiles

Materials called geotextiles are applied to soil to aid in drainage, erosion control, and soil stability. Geotextiles are available in various thicknesses and sizes. Some are made of woven materials, while others are four inches thick plastic extrusions. The principal purpose of geotextiles is to stop soil erosion. They prevent water or rain from moving the top layer of soil.

In certain situations and for certain uses, geotextiles can be utilized in the construction of roads. For road construction, thicker extrusion geotextiles are usually needed to give the necessary strength. Geotextiles can be applied to practically any type of soil and are available in a range of sizes and shapes. They are rarely utilized for large, heavy-load applications or road construction due to their high cost. Because workers must manually lay them out, geotextiles are usually labor-intensive due to their mechanical nature.

Advantages

Some advantages of soil stabilization using geotextiles are:

- Long-term, permanent
- Most soil types are compatible with it
- Various types for different applications

Disadvantages

- Time-consuming
- Labor intensive
- Expensive

2.3.4.7. Bituminous Soil Stabilization

Bituminous Soil Stabilization frequently uses bituminous materials like tar and asphalt. Although bituminous stabilization lowers the quality of water absorption, it increases soil stability. This behavior prevents the soil from eroding quickly, preserving the surface for a longer amount of time. In addition to bitumen, mixtures of soil, sand, oil, and other materials are used for better outcomes (Al-Tabbaa and Evans, 2005; EuroSoilStab, 2002; Porbaha et al., 2005).

2.3.5. Some Other Methods of Soil Stabilization

Despite the above-mentioned soil stabilization methods, numerous other methods are sometimes used if the above methods do not apply to the situation.

Other methods of soil stabilization include:

2.3.5.1. Mass Stabilization Method

A shallow-to-deep stabilization technique called mass stabilization can stabilize all of the soft soil up to a predetermined depth. Although the method is still in its infancy, it works well for stabilizing soils with high moisture content, including silt, clay, organic soils, and contaminated sediments. The soil mass can be blended by self-injecting the binder into a rotating auger or mixing head and the soil, or by using an excavator-mounted mixing tool with special shuttles that pneumatically deliver the binder to the tool's head and into the mix zone.

When it comes to site remediation, mass stabilization is an affordable way to improve the ground, particularly in cases where there are a lot of contaminants and high-water content.

2.3.5.2. Deep Mixing Method

The deep mixing method involves stabilizing soils at great depths. It is an in-situ ground modification technology in which a wet or dry binder is injected into the ground and mechanically or rotary mixed with in-situ soft soils (clay, peat, or organic soils).

2.3.5.3. Grouting Method

In the grouting method of soil stabilization, stabilizers are injected into the soil to stabilize the soil. This method is ineffective for clayey soils as they have low permeability. This is an expensive method of soil stabilization. The grouting method is appropriate for stabilizing buried zones of relatively limited size (Figure 2.3).

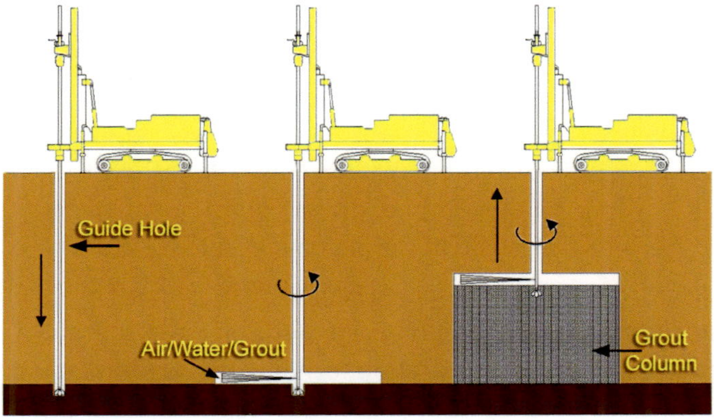

Figure 2.3. *Grouting method of soil stabilization.*

Source: By U.S. Army Corps of Engineers Sacramento District – Illustration of jet-grout approach to precisely inject seepage cutoff walls inside levees, Public Domain, https://commons.wikimedia.org/w/index.php?curid=52214209.

2.3.5.4. Electrical Method

The process of applying an electrical current to a soil mass (under consideration) in order to move charged soil particles around and alter the soil's characteristics, ultimately leading to overall stabilization of the soil, is known as electrical stabilization. The weak clayey soils are the main application for this soil stabilization technique. Weak clayey soil, as we all know, requires high-strength treatment. Additionally, because of their low hydraulic conductivity, chemical grout cannot be used with conventional hydraulic methods.

2.3.6. Environmental and Contamination Considerations

Two important factors in soil stabilization are soil contamination and environmental factors. Heavy metals, organic pollutants, and industrial byproducts are examples of contaminants that can negatively impact the effectiveness of stabilization methods and necessitate the use of specific remediation techniques. For example, before stabilization, contaminated soils might need to be treated to eliminate or neutralize dangerous materials.

The efficiency of stabilization treatments can also be impacted by external factors like exposure to chemicals, seasonal variations in moisture, and extremely high or low temperatures. For soil stabilization techniques to be effective and durable over the long term, these factors need to be taken into account during selection and application.

2.3.7. Testing and Quality Control

During the entire soil stabilization process, continuous testing and quality assurance are crucial. To make sure the soil fulfills the necessary requirements and functions as intended, it is important to regularly monitor the properties of the soil, the quality of the compaction, and the efficacy of stabilization treatments. Real-time data on soil conditions during construction is provided by field tests like moisture content measurements and in-situ compaction tests.

Particle size distribution tests, plasticity tests, strength tests, and other laboratory analyses provide comprehensive insights into soil behavior and aid in the selection of suitable stabilization techniques. Performance assessments, like load tests and erosion tests, determine any problems that need to be fixed and evaluate how well stabilization treatments work.

2.3.8. Case Studies and Practical Applications

As an example of how soil stabilization techniques are used in practice, take a look at a few case studies that use various strategies. For instance, lime stabilization has been utilized in road construction projects to enhance the performance of clayey soils, leading to an increase in load-bearing capacity and a decrease in maintenance needs. On the other hand, in urban development projects, cement stabilization has been utilized to increase the durability and strength of soil that is used to build foundations.

Biological stabilization techniques have been effectively applied to improve slope stability and stop soil erosion in coastal areas. Furthermore, to address environmental concerns and enhance soil performance in areas with contaminated soils, stabilization techniques and remediation techniques have been combined. Assessing the soil's suitability for stabilization entails a thorough analysis of the soil's composition, moisture content, compaction properties, and surrounding circumstances.

Through meticulous evaluation of these variables and the selection of suitable stabilization techniques, engineers can optimize soil performance and guarantee the triumph of construction endeavors. In order to ensure the safety, longevity, and functionality of infrastructure, soil stabilization is a crucial component of civil engineering. For this reason, it is crucial to use a comprehensive and knowledgeable approach to soil evaluation and treatment (Al-Tabbaa and Evans, 2005; EuroSoilStab, 2002; Porbaha et al., 2005).

REFERENCES

1. Al-Tabbaa, A., & Evans, W. C. (2005). Stabilization-solidification treatment and remediation: Part I: Binders and technologies—Basic principles. In *Proceedings of the International Conference on Stabilization/Solidification Treatment and Remediation* (pp. 367–385). Cambridge, UK: Balkema.
2. Beeghly, J. (2003). Recent experiences with lime-fly stabilization of pavement subgrade soils, base, and recycled asphalt. In *International Ash Utilization Symposium* (Paper No. 46). University of Kentucky.
3. EuroSoilStab. (2002). *Development of Design and Construction Methods to Stabilize Soft Organic Soils: Design Guide for Soft Soil Stabilization* (CT97-0351). European Commission, Industrial and Materials Technologies Programme (RITE-EURAM III).
4. Hogentogler, C. A., & Terzaghi, K. (1929). Interrelationship of load, road, and subgrade. *Public Roads*, 37–64.
5. Holm, G., Andréasson, B., Bengtsson, P., Bodare, A., & Eriksson, H. (2002). Mitigation of track and ground vibrations by high-speed trains at Ledsgård, Sweden. Linköping: Swedish Deep Stabilization Research Centre.
6. MacLaren, D. C., & White, M. A. (2003). Cement: Its chemistry and properties. *Journal of Chemical Education, 80*(6), 623.
7. Maher, A., Bennert, T., Jafari, F., Douglas, W. S., & Gucunski, N. (2004). Geotechnical properties of stabilized dredged material from New York-New Jersey Harbor. *Journal of the Transportation Research Board*, 86–96.
8. Porbaha, A., Shima, M., Miura, H., & Ishikura, K. (1999). Dry jet mixing method for liquefaction remediation. In *Proceedings of the International Conference on Dry Mix Methods of Deep Soil Stabilization*. Stockholm: Balkema.
9. Porbaha, A., Weatherby, D., Macnab, A., Lambrechts, J., Burke, G., Yang, D., & Puppala, A. J. (2005). American practice of deep mixing technology. In *Proceedings of the International Conference on Deep Mixing—Best Practice and Recent Advances*. Stockholm.
10. Stocker, M., & Seidel, A. (2005). Twenty-seven years of soil mixing in Germany: The Bauer mixed-in-place technique. In *Proceedings of the International Conference on Deep Mixing—Best Practice and Recent Advances*. Stockholm.

CHAPTER 3

SOIL CLASSIFICATION AND TESTING FOR STABILIZATION

Contents

INTRODUCTION

Chemical stabilization techniques employ chemical additives to enhance the soil's engineering qualities, making it more appropriate for construction endeavors, especially those involving the building of roads, the support of foundations, and the management of erosion. In addition to mitigating problems like swelling and shrinkage, which can cause structural instability, the procedure increases the strength of the soil and decreases its permeability. As an alternative to the more costly and time-consuming soil replacement process, chemical stabilization is frequently used when the current soil conditions are unsuitable for construction.

3.1. LIME STABILIZATION

The process of enhancing the qualities of soil by a variety of techniques, such as compaction, blending with chemicals or additives, or other non-destructive techniques, to increase the soil's strength and durability in addition to other benefits, is known as soil stabilization. The two main types of soil stabilizers are conventional and non-traditional; the former include fly ash, cement, and lime; the latter include sulfonated oils, asphalt, and ionic stabilizers. For each of these, there are significant variations in the soil-stabilizer mechanism as well as the properties following stabilization. These variations are caused by the soil's heterogeneity, variations in micro- and macrostructure, variations in geologic deposits, and variations in the interactions between the soil-stabilizer mix's chemicals and physical components. Geotechnical engineers prefer lime and cement as stabilizers because of their ease of mixing, cost-effectiveness, and overall property improvement. Lime is especially useful for treating problematic expansive soils.

Lime stabilization of soil is not a novel technique; the Chinese and Indians have long employed this specific mixture for underground chambers, bridge footings, and dam construction. Road sub-bases in Rome were made of a mixture of soil and lime. They did not, however, have any written protocols or guidelines. However, it wasn't until after 1945 that lime stabilization with precise guidelines and a defined process began. Subsequently, this stabilization technique was widely applied in building foundations using lime columns, subgrade improvement, embankment stability, bridge abutments, and sliding slope replacement material.

Lime stabilization is the process of stabilizing soil by applying either hydrated lime ($Ca(OH)_2$) or quick lime (CaO). This method works best on fine-grained soils with a high percentage of clay. Quick lime is more frequently chosen due to its potency and faster development of strength. Bell claims that clays with higher silica contents react more with lime; to put it another way, 3-layered clay minerals, like montmorillonite, have two sides that expose silica, making them more reactive than 2-layered minerals, like illite, which only expose one side of the mineral to silica. When lime is applied to clay-rich soils, the liquid limit, maximum dry density (MDD), and plasticity index decrease while the optimum moisture content (OMC), strength, and shrinkage limit increase. For soil modification, 1 to 3% of lime is typically required, i.e., reduction in the soil's plasticity and 2 to 8% are necessary for real stabilization, i.e., cementation. The percentage of lime needed for total soil modification and stabilization is known as the optimal lime content, i.e., lime content, which raises strength, durability, and stiffness while also improving workability and reducing plasticity clayey soils with varying degrees of plasticity, i.e., lime stabilization works best with PI > 15 (Doty and Alexander, 1968; Winterkorn, 1975; Zuber et al., 2013).

3.1.1. Mechanism of Lime Stabilization

According to Eades and Grim, the lime stabilization procedure is basically a two-stage procedure. The reactions occurring between clay and lime were identified using differential thermal analysis (DTA) and X-ray diffraction. Depending on the mineralogical makeup of the soil, Cation Exchange, Flocculation, Carbonation, and certain short-term pozzolanic reactions also take place during the first stage of modification, which can take several hours or days to complete. At this point, the soil's workability and plasticity are mostly altered. The pozzolanic reactions that contribute to the improvements in strength and durability occur during the second stage, which is also referred to as long-term treatment.

3.1.1.1. Short-Term Treatment

The calcium ions replace monovalent ions like Na^+ and Li^+ when water and lime are added to the clayey soil, causing the lime to dissociate into Ca^{2+} and OH^-. This process is also referred to as a cation exchange (CEC) process, etc., which are adsorbed to the clay mineral surface. The CEC is generally determined by the clay mineralogy and pH of the soil water; clays rich in montmorillonite have the highest CEC and clays low in kaolinite have the lowest CEC. The process known as flocculation occurs when Ca^{2+} ions attach to clay minerals, causing a net decrease in repulsive forces as well as a reduction in the thickness of the diffused water layer. As a result, soil particles become closer to one another. Little and Nair claim that this flocculation process produces large-sized particle aggregates, less plastic, and soils that are more friable and workable. SEM micrographs confirmed that the structure of the dredged da li soil changed from a deflocculated to a flocculated state upon treatment with lime.

3.1.1.2. Long Term Treatment

The pH rises as a result of the lime dissociating and producing OH^-. Cementitious hydrates are created when the pH level rises above 12.4, causing the silicate and alumina in the clay minerals to dissolve and separate from the soil. They then combine with the calcium ions that are created when lime hydrates. e. two types of hydrates are calcium aluminate (C-A-H) and calcium silicate (C-S-H). The following are the chemical processes that go into pozzolanic cementation.

$$Ca(OH)_2 + SiO_2 \longrightarrow CaO\text{-}SiO_2\text{-}H_2O$$

$$Ca(OH)_2 + Al_2O_3 \longrightarrow CaO\text{-}Al_2O_3\text{-}H_2O$$

According to Brandl, there is no increase in strength after seven years; the strength improvement rate declines with time and is greatest in the first two years. He added that a strength increase that is time-dependent increases linearly with the logarithm of time. Temperature, pH level, calcium content, and the amount of silica and alumina in soil minerals all affect pozzolanic reactions. According to Little and Nair, the pozzolanic reaction is a cooperative mechanism between clay and lime in which the former creates a high pH environment that solubilizes alumina and silica and the latter provides calcium-free ions that form cementitious materials when combined with alumina and silica (Doty and Alexander, 1968; Winterkorn, 1975; Zuber et al., 2013).

3.1.2. Changes in Soil Properties Due to Lime Stabilization

When fine-grained soils are treated with lime, there is a noticeable change in the soil's permeability, limits on swell and shrinkage potential, and overall strength increases. However, the quality of the lime, the type of soil, time, temperature, and moisture all affect how the physical properties change. Temperature and time have a major impact on strength, pH, and reactivity. According to Thompson, there are two types of soil treated with lime: modified and reactive. The UCS value of the modified soils increases by less than 345 kPa, while the UCS value of the reactive soils increases by more than 345 kPa. These soils exhibit notable changes in their overall workability, swell potential, plasticity, and soil texture. The quantitative and qualitative examination of the modifications in soil properties brought about by lime treatment is provided below:

3.1.2.1. Effect of Lime Treatment on Plasticity

When soil is treated with lime, its plasticity is significantly reduced, and it frequently stops being plastic altogether. The decrease in the liquid limit and the increase in the plastic limit are the causes of this reduction. However, Croft explained the increase in the liquid limit that Zolkov had noticed as the result of hydroxyl ions present altering the clay surface's affinity for water. According to Dash and Hussain, applying lime to the soil results in two main effects: first, a decrease in the liquid limit because the diffuse double layer thickness is reduced; second, changes in the soil's structure lead to an increase in the liquid limit because pozzolanic reactions produce water-holding gelatinous products. They also concluded that when the diffuse double layer's thickness decreases, charge concentration rises and the pore fluid's viscosity rises as a result. This raises the shear resistance between the particles and results in a net increase in the plastic limit. The type of soil determines whether the liquid limit rises or falls. Bell applied a lime treatment to three different soils with the primary minerals being montmorillonite, quartz, and kaolinite. He found that the soils with montmorillonite as the principal clay mineral usually had the biggest increase in the plastic limit. The amount of lime needed to reach the non-plastic stage is referred to as the lime fixation point. If the plasticity index and clay content are higher, then more lime is needed to achieve non-plastic conditions. Usually, 1 to 3% of lime is needed to alter the soil, i.e., quick lime seems to be more beneficial than hydrated lime in terms of improving shrinkage characteristics when it comes to plasticity changes. Additionally, the texture of the soil changes significantly. More friable soil is more workable, which facilitates easier soil placement.

Clayey soils treated with lime also experience a reduction in moisture absorption due to an increase in swell potential. Due to a higher electrolytic concentration and a thinner diffuse double layer, Dash and Hussain observed a significant initial decrease in swell potential upon adding lime. However, additional lime addition resulted in an increase in swelling because of silica gel formation, a phenomenon that is more common in silica-rich soils. Seed, Woodward and Lundgren developed a relation between the plasticity index and swell potential which states:

$$\text{Swell Potential} = 0.00216 \times \mathbf{PI}^{2.44}$$

where: PI is the Plasticity Index.

3.1.2.2. Effect of Lime Treatment on Moisture Density Relations

Grain size increases as a result of cation exchange and the flocculation process, which also raises the void ratio and lowers the MDD while raising the ideal moisture content. It is feasible to reach the required density over a larger range of moisture content because the compaction curve flattens after a lime treatment. Jan and Mir associate these modifications with the enhanced cation concentration-induced decrease in the diffuse double layer's thickness. Bell came to the conclusion that clay rich in montmorillonite exhibits the greatest variations in ideal moisture content and dry density, and that compaction curve distortion may even occur. He added that the amount of time that passes between mixing and compaction determines how much density decreases. The ideal moisture content is typically increased by 2 to 4% after lime treatment, and the dry density is decreased by 48 to 80 kg/m^3.

3.1.2.3. Effect of Lime Treatment on Strength

The effects of lime on soil strength have been confirmed by researchers all over the world using triaxial testing, California Bearing Ratio (CBR), and Unconfined Compressive Strength (UCS). As pozzolanic products that have a cementitious effect on soil particles develop, lime-treated clays typically exhibit significant strength gains. Thompson et al., report that after 28 days of curing, the UCS strength increased by about 60% as a result of pozzolanic reactions. The characteristics of clay mineralogy, soil pH, silica-alumina percentage, lime type, moisture content, temperature, and curing time are among those that have a significant impact on strength gain. Compared to soils with low clay content, those with a higher clay fraction need more lime to gain strength because they have more lime available for pozzolanic reactions and a lower percentage of lime needed for plasticity amendment. Strength gain is thought to be significantly influenced by curing; after the first seven days, strength gain accelerates and then slows down. Eades and Grim concluded that soil mineralogy and lime percentage are the primary governing factors of strength gain after performing UCS on six soils with varying mineralogies. Compared to an alkaline environment, the acidic environment's lime stabilization produced less UCS change. Low plasticity California soils (clayey silts and silty clays) showed a significant strength gain after lime treatment, suggesting that a small amount of reactive clay may be enough to produce a significant strength gain.

The strength of a material tends to decrease as the lime content increases beyond a certain point. This is because an excess of Ca^{2+} ions can cause clay particles to deflocculate and split, which can result in cracks even at low loads and a lower UCS value. According to Bell, the cohesion and friction of lime are negligible, and adding lime beyond an ideal level merely lubricates the soil particles, which reduces strength. Together with a slight increase in the angle of internal friction, fine-grained clay soils exhibit a significant increase in cohesiveness. Thompson related cohesion and Unconfined compressive strength (UCS) with the following expression:

$$C = 9.3 + 0.292\sigma_c$$

Where C is cohesion(psi) and σ_c is unconfined compressive strength (psi).

To look for improvements in the strength parameters, undrained triaxial tests were performed on plastic Jordanian clay. The results showed a significant increase in the undrained cohesion and angle of internal friction, which was linked to the amount of lime added and the curing period. After performing Direct Shear tests on both treated and untreated dredged soil, Jan and Mir found that cohesion and angle of internal friction increase significantly up to an optimal lime content of 8% and then show a downward trend. Mateos noted that soil treated with lime at 35°C developed twice the strength of the same soil specimen treated at 25°C, suggesting that higher temperatures hasten the curing process and result in stronger soil. The pozzolanic reactions between lime and soil slow down and may even stop below 4°C. Because of the lime treatment, tensile strength increases along with UCS. For the purposes of design, the modulus of rupture to UCS is 0.25, and the ratio of tensile strength to UCS is roughly 0.13.

3.1.2.4. Effect of Lime Treatment on Compressibility

One of the key elements influencing the stability of structures like buildings and roads is the compressibility of the soil. Several investigations have been conducted to determine how lime treatment affects the properties of soil compressibility. Rajasekaran and Rao examined the impact of lime treatment on the compressibility of marine clay using the injection technique and lime column work. After 30 to 45 days, consolidation tests were conducted. The results showed a significant improvement in soil incompressibility, with half to one-third of the untreated soil showing a decrease in the compression index from 0.85 to 0.36. Up until a certain point of lime addition, the resistance against compression increases; after that, no increase takes place.

Tedesco conducted several experiments and concluded that pozzolanic reactions had little effect and that short-term reactions were the cause of the decrease in compressibility of soil treated with lime. He added that when compaction is carried out on the dry side of optimal, lime-soil's compressibility decreases.

3.1.2.5. Effect of Lime Treatment on Durability and Fatigue Strength

Soils are frequently exposed to alternating cycles of wet-dry and freeze-thaw, which negatively impacts soil strength and stability. The capacity of soil to tolerate these impacts is referred to as durability. Mallela et al., recommend using the cyclic freeze-thaw test and the soaked strength test to evaluate the longevity of lime-treated soil. Thompson tested the compressive strength of lime-treated soil both soaked and unsoaked, and found that the ratio of soaked to unsoaked strength ranges from 0.7 to 0.85. Because the lime-soil mix's autogenous healing helps the soil recover its stability following winter freeze-thaw cycles, the distress is not cumulative. Brandl concluded that the lime-soil mix has low heat conductivity because of the higher void ratio caused by flocculation, which lowers the depth and rate of frost penetration. Mallela et al., say that the number of load cycles the material can withstand at constant stress is indicated by its fatigue strength. Because of the steady development of strength over time, the fatigue strength of the lime-soil mix is increased (Doty and Alexander, 1968; Winterkorn, 1975; Zuber et al., 2013).

3.1.3. Applications of Lime Treatment

Calcium-based stabilizers, such as lime, are used for several reasons: (i) replacing fine-grained problematic soils with coarse-grained soils is often costly; (ii) the presence of Ca^{2+} accelerates pozzolanic reactions and reduces swell potential; (iii) alternative techniques, such as pre-wetting, are time-consuming and thus unsuitable; and (iv) environmental and economic benefits are associated with this type of soil stabilization. Currently, lime treatment of soils is widely used as a primary method of soil stabilization, and advanced technologies are being developed to refine lime treatment techniques across various fields. Some of these areas and techniques are discussed below.

3.1.3.1. Lime Columns

According to Broms, utilizing specialized augers, soil can be mixed *in-situ* with lime to improve the behavior of soft clays. It was first thought that the stabilized soil column supported the superstructure by acting as a pile, but additional research revealed that lime can diffuse into the surrounding soil and stabilize a bigger volume of soil. Even in highly sensitive clays, according to Broms, the shear strength of stabilized soft clay will surpass that of untreated soil approximately one to two hours after mixing. If the initial shear strength is 10 to 15 kPa, an increase of 10 to 50 times is anticipated after a year. Additionally, he concluded that the lime column stabilizes soil in three stages. When the quick lime absorbs water from the soil, it first causes slaking and swelling of the lime, which consolidates the surrounding soil.

$$CaO + H_2O \longrightarrow Ca(OH)_2 + 15.6\ \text{Kcal/mol}$$

Ekstrom and Trank found that after treating soft clay with 10% lime, the water content dropped from 60% to 45%. Following the first consolidation, the Ion Exchange process begins, causing the clay surface to absorb the Ca^{++} ions from the slaked lime. As a result, the clay particles bind together, leading to a net increase in shear strength. Ultimately, Pozzolanic reactions occur and play a part in the long-term increase in shear strength. When Brandl performed triaxial tests at low cell pressures, he found that the amount of lime increased along with the angle of internal friction. According to Bell and Tyrer, stabilized soil is less plastic, has a lower sensitivity (about 1 to 3) and is less compressible than untreated soil. Following treatment, the soil's texture becomes granular and is typically firm to hard; it also acts like an overly-consolidated clay.

The deep mixing method, developed in Japan, employs multiple mixing units mounted on a single barge. These mixers are closely spaced and exhibit greater efficiency. In countries like India, China, and Malaysia, predrilled holes are filled with unslaked lime. Over time, the soil between boreholes consolidates due to the expansion caused by the slaking of lime. Figure 3.1 provides a pictorial representation of the elements and processes involved in lime column manufacture, while Figure 3.2 depicts a lime-column installation rig typically used for large-scale soil stabilization.

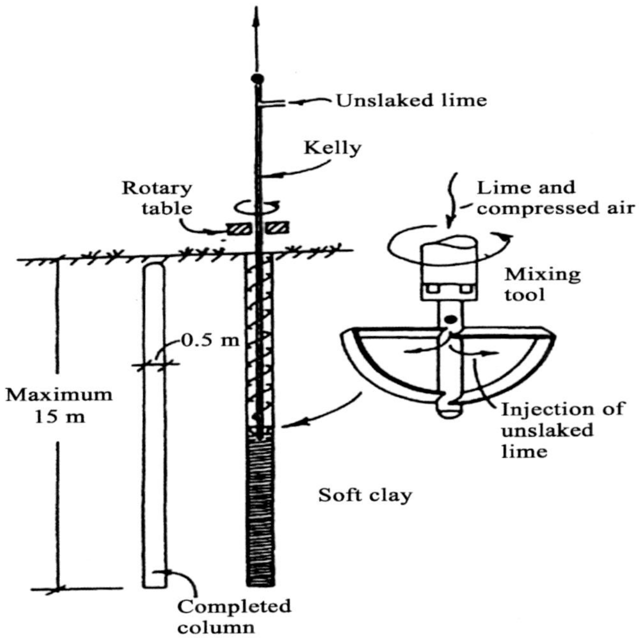

Figure 3.1. *Manufacture of lime columns.*

Source: https://www.jmaterenvironsci.com/Document/vol11/vol11_N9/JMES-2020-11133-Manzoor.pdf.

Figure 3.2. *Lime column rig.*

Source: https://www.jmaterenvironsci.com/Document/vol11/vol11_N9/JMES-2020-11133-Manzoor.pdf.

3.1.3.2. Pavement Stabilization

For many years, base courses and subgrades have been stabilized by adding lime to improve various aspects of their properties, including their ability to support loads. The term "resilient modulus" refers to a property of soil that is measured through dynamic testing in which a uniaxial compressive load is applied. The resilient modulus of soil subgrades and subbases stabilized with lime was assessed by Little using a Falling Weight Deflectometer (FWD) and Maxwell and Joseph using a field vibration test. According to Little and Nair, the reduction of plasticity and pozzolanic reactions following lime treatment increase the cohesive strength of soils and aggregate systems, which has a significant stiffening effect as indicated by the resilient modulus magnitude. When lime is applied and the soil is prepared with a high moisture content and low densification level, Fossberg reported a 20-fold increase in the resilient modulus of montmorillonite cohesive soils. Figure 3.3 illustrates how to stabilize the subgrade with lime in the field using large machinery for mixing and compaction.

Figure 3.3. *Stabilization with lime.*

Source: Połomski M, Wiatkowski M. (2022). The use of lime for drainage of cohesive soils built into hydraulic engineering embankments. Water, 14(22):3700. https://doi.org/10.3390/w14223700.

Little evaluated the resilient modulus of the sub-bases of six pavement sections close to Phoenix, Arizona, using FWD; these sections all had comparable cross-sections and construction materials. Nevertheless, 1% lime was used to stabilize three of these subbases, leaving the other three unstabilized.

It is clear from the above table that subbases treated with lime have a significantly higher resilient modulus and fatigue life than unstabilized subbases. Additionally, because stabilized subbases have better load-spreading capabilities, the compressive subgrade strains and tensile flexural strains are significantly lower. Nowlin and associates utilized FWD to back calculate the resilient modulus of 12 pavement sections with different subgrade soil properties throughout Texas State that had lime-stabilized soil subgrades. He found that each subgrade's unique soil properties had a significant impact on the amount of resilient modulus that increased after lime treatment; some subgrades showed notable improvements, while other low plastic soils saw a net decrease in resilient modulus. These tests were conducted after a considerable amount of time, and the increase in stiffness attests to the longevity and durability of the structural alterations brought about by lime stabilization. Using highly plastic Burleson clay stabilized with 5% lime and no pavement layers, little carried out road tests. Resilient modulus was measured with FWD both before and after trafficking pavement with 5,000 applications of an 80 kN axle load. Based on the findings, he deduced that lime stabilization causes a notable rise in the resilient modulus of clay subgrade, which persisted even after two years of exposure to heavy rainfall and 5,000 applications of an 80 kN single axle load. After analyzing the data, Peterfalvi et al., recommended a lime-stabilized subgrade depth of 0.25 to 0.35 meters for good long-term performance of road pavements on cohesive soils. They also built three experimental roads to determine the bearing capacity of the stabilized soil layer and to study traffic resistance of roads built on stabilized soil (Table 3.1).

Table 3.1. Chemical Characteristics of Quicklime as a Building Material

Component	Content
CaO + MgO	≥90%
MgO	≤5%
CO_2	≤4%
SO_3	≤2%
Available lime	≥80%

Source: Połomski M, Wiatkowski M. (2022). The use of lime for drainage of cohesive soils built into hydraulic engineering embankments. Water, 14(22):3700. https://doi.org/10.3390/w14223700.

3.1.3.3. Lime Stabilization for Dams and Slopes

Around the world, especially in developed nations, lime stabilization has been successfully utilized to lessen the vulnerability of dams and slopes to erosion, piping failures, and slip surface failures; in this sense, many studies have been conducted. Constructed in 1951 using montmorillonite clay, the Friant-Kern canal sustained damage in several places and was restored with a soil lime mixture in 1973–1977. The effectiveness of lime treatment was demonstrated by the researchers' observation that, as a result of the treatment, very few maintenance procedures were needed, no failures or slides were reported, and only very slight surface erosion occurred in some areas. Additionally, lime was applied to more than 150 Mississippi River Levee failure sites to stabilize the slopes, with successful outcomes.

Herrier et al. (2012) carried out a thorough investigation to determine the most effective methods for applying lime treatment in hydraulic structures and to examine the impact of lime on soil characteristics. After doing experiments, they found that lime treatment reduced permeability because it reduced pore size, as shown by mercury intrusion porosimetry. They associated high moisture content and kneading compaction with the achievement of low permeability.

They also reported a 25-fold increase in cohesiveness and a 5–10-fold decrease in swell potential, which they linked to particle adhesion and cementation. Tests for erosion resistance were carried out both internally and externally, and notable advancements were noted. They came to the conclusion that lime stabilization reduced permeability and improved mechanical properties, volume stability, and erosion resistance—all of which are crucial elements that affect the durability, design, and construction of hydraulic structures.

3.2. CEMENT STABILIZATION

The hydration of the cement particles causes the soil particles to bond together and grow into crystals that can interlock to give the soil a high compressive strength. This process is known as soil cement stabilization. The majority of the material particles must be coated by the cement particles for a successful bond to form. To achieve effective stabilization of soil cement through good contact between soil particles and cement, a specific particle size distribution must be mixed with the cement and soil.

3.2.1. Advantages of Cement Stabilization

While several reagents can be used for Soil Stabilization, Portland cement has advantages that make it more economical and easy to use than others:

- Cement is manufactured under strict ASTM standards, ensuring uniformity of quality and performance
- Cement's success in Soil Stabilization is supported by more than 50 years of use on a variety of projects
- Cement has a long-term performance record
- Using cement can minimize volume increase compared with other reagents
- Cement is a non-proprietary manufactured product, readily available across the country in bag or bulk quantities.

3.2.2. Problems Associated with Cement Stabilization

Cement stabilized materials have certain drawbacks despite their many advantages, which should be taken into account. Cracking and carbonation are the main issues that, if left unchecked, will have noticeable negative effects. These issues arise in the stabilized, compacted layer following construction.

The causes of cracking in cement-stabilized bases include subgrade constraints on the base, construction methods, traffic loads, and material properties. Shrinkage cracks are the most prevalent kind of cracks in bases stabilized with cement. The causes of shrinkage cracks include water loss, cement concentration, compacted material density, compaction technique, and the moisture content of the material to be stabilized before treatment. If proper curing is not exercised, cement-treated materials start to lose moisture through evaporation immediately after they are placed. The drying process that follows the moisture loss will cause shrinkage cracks to appear. Furthermore, as a result of the cement's hydration being hindered by insufficient moisture in the mix, the final strength of the cement-treated materials will be decreased.

The contribution of cement hydration in the development of shrinkage cracks is less as compared to water loss. Nevertheless, an excessive amount of cement aggravates the development of cracks in two ways:

- A higher cement content in the mixture results in a higher hydration water consumption, which raises the drying shrinkage.
- The treated materials become more rigid and have a higher tensile strength when the amount of cement increases. Wide cracks that are widely spaced develop as a result. The higher tensile strength is responsible for the wider crack spacing, while the distribution of the material's total shrinkage within fewer, widely spaced cracks is the cause of the wider width of individual cracks.

3.2.3. Cement Stabilized Subgrade Construction Procedure

Construction of cement-stabilized subgrade soil follows this procedure-

- Subgrade material was spread on the top of the embankment layer (150 mm + 25% loose) and lightly compacted.
- Total work area marked with grids of the area required for one bag of cement (2.5 m × 2.5 m).
- Cement spread at the rate of one bag per grid.
- Soil and cement are mixed by mechanical harrow followed by tractor-mounted rotovator.
- Moisture added to the soil (+/-1% of OMC).
- Quality Control tests conducted as per the frequency.

The methodology for cement stabilized subgrade can be described with the help of pictures taken at the site (Figure 3.4).

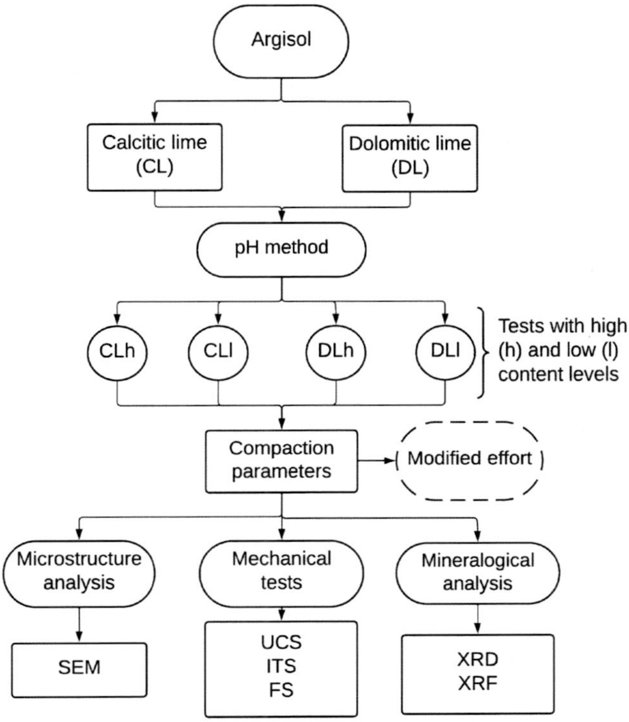

Figure 3.4. *Showing cement stabilized subgrade at a test site.*

Source: Diniz BC, Fedrigo W, Kleinert TR, Batista GdS, Núñez WP, Correa BM, Brito LAT. (2024). Lime stabilization of tropical soil for resilient pavements: mechanical, microscopic, and mineralogical characteristics. Materials, 17(19):4720. https://doi.org/10.3390/ma17194720.

3.2.4. Applications of Cement Stabilization

Many applications in civil engineering call for the stabilization of cement, especially those involving weak or unstable soils that must be strengthened to support building projects. Some common applications include:

1. **Road and Pavement Construction:** When building roads and pavements, cement stabilization is commonly used to enhance the base layers and subgrade. Cement stabilization contributes to the long-term performance of the pavement by strengthening and extending these layers' resilience, which lessens the need for upkeep and repairs.

2. **Foundation Stabilization:** Cement stabilization can be applied to weak or expansive soil areas to increase the soil's bearing capacity and suitability for supporting building foundations. This is especially crucial in areas with clayey soils, as foundation stability can be a major issue.
3. **Embankment Construction:** Additionally, cement stabilization is used to build embankments, where it contributes to the soil's increased strength and stability. This is especially crucial in places where landslides or erosion are common, as the stability of the embankment is essential to the structure's longevity and safety.
4. **Slope Stabilization:** Cement stabilization is an additional tool for preventing landslides and stabilizing slopes in addition to embankments. Cement stabilization lowers the risk of slope failure by strengthening and uniting the soil, and safeguarding natural landscapes and infrastructure.
5. **Erosion Control:** Cement stabilization is a useful component of erosion control strategies, especially in regions where soil erosion is caused by wind or water. Cement stabilization aids in preventing soil loss and preserving the integrity of the landscape by binding the soil particles together.
6. **Soil Improvement for Agriculture:** Cement stabilization is a technique that can sometimes be used to enhance the fertility or drainage of agricultural soils. Through cement stabilization, agricultural systems can become more productive and sustainable by enhancing the soil's structure and stability.

3.2.5. Challenges and Limitations of Cement Stabilization

Cement stabilization has many advantages, but it also has drawbacks and restrictions. Making decisions regarding the use of cement as a soil stabilizer and guaranteeing the success of stabilization projects depend on having a clear understanding of these difficulties.

- **Soil Variability:** The variability of soil properties, which can impact the performance of the stabilized material, is one of the main obstacles to cement stabilization. It is possible for different soils to respond differently to the stabilization process or to require varying amounts of cement. To guarantee the success of the stabilization project and determine the proper cement content, thorough testing and analysis are required.
- **Environmental Impact:** Cement production is linked to significant

carbon dioxide emissions, even though the stabilization of cement can have positive effects on the environment in certain situations. When selecting cement stabilization, one must take the environment into account, especially in projects with stringent sustainability requirements. The environmental impact of cement stabilization can be minimized by using supplementary cementitious materials (SCMs) such as slag or fly ash, or by using alternative stabilization techniques.

- **Moisture Sensitivity:** Soils stabilized with cement may be moisture-sensitive, especially while they are curing. While too little moisture can hinder the cement's ability to properly hydrate, too much moisture can cause problems like diminished strength and durability. In cement stabilization projects, careful moisture content control is crucial to getting the best possible results.
- **Time-Dependent Strength Gain:** Over time, as the hydration and pozzolanic reactions advance, cement-stabilized soils become stronger. Although this is usually a good thing, it can also be a drawback if construction requires instant strength. In certain instances, the gradual increase in strength might require the utilization of extra materials or methods to attain the intended results within the project schedule.
- **Cost Considerations:** While cement stabilization is generally cost-effective, the cost of cement can be a limiting factor in some projects, particularly in regions where cement is expensive or difficult to obtain. Additionally, the cost of equipment and labor for mixing and applying cement may add to the overall expense of the stabilization process.
- **Durability Concerns:** While cement stabilization increases soil durability, other factors like sulfate attack, freeze-thaw cycles, and the presence of aggressive chemicals in the soil or groundwater can affect the long-term performance of cement-stabilized materials. It is imperative to tackle these durability issues utilizing meticulous material selection, appropriate design, and suitable construction techniques (Doty and Alexander, 1968; Winterkorn, 1975; Zuber et al., 2013).

3.3. FLY ASH STABILIZATION

A well-known technique for improving the engineering qualities of soil is fly ash stabilization, especially when it comes to building roads and developing land. The main constituents of fly ash, a byproduct of burning coal in thermal power plants, are fine particles that are high in silica, alumina, and calcium oxide. Because of its affordability, favorable effects on the environment, and capacity to increase soil strength, durability, and load-bearing capacity, its application in soil stabilization has grown in popularity.

3.3.1. Properties of Fly Ash

Fly ash is distinguished by its pozzolanic qualities and fine, powdery texture. Materials classified as pozzolanic react with calcium hydroxide in the presence of water to produce compounds with cementitious qualities. Fly ash's ability to create bonds between soil particles, which boosts the soil's strength and stability, makes it a useful stabilizer. Depending on the kind of coal burned, fly ash's chemical makeup varies, but it typically includes significant amounts of silica (SiO_2), alumina (Al_2O_3), and calcium oxide (CaO), along with trace amounts of iron oxide, magnesium oxide, and other elements.

3.3.2. Mechanisms of Fly Ash Stabilization

Physical and chemical processes work together to stabilize soil when fly ash is used. In terms of physical properties, the fly ash fills in the spaces between soil particles, decreasing the permeability and raising the density of the soil. The soil structure becomes more stable and compacted as a result. In the presence of calcium hydroxide, fly ash and water undergo a pozzolanic reaction that produces cementitious materials like calcium silicate hydrate (CSH) and calcium aluminate hydrate (CAH). By binding the soil particles together, these substances improve the soil's ability to support weight and resist deformation.

3.3.3. Types of Fly Ash

Fly ash comes in two primary varieties: Class C and Class F. Class C fly ash, sometimes referred to as high-calcium fly ash, has self-cementing and pozzolanic qualities and contains more than 20% calcium oxide (CaO). This means that Class C fly ash is especially useful for stabilizing soil because it can react with water alone to form cementitious compounds. In contrast, Class F fly ash is mostly pozzolanic and has less than 20% calcium oxide. To initiate the pozzolanic reaction, a cementing agent like lime must be added. Depending on the particular soil conditions and intended results, either type of fly ash can be used to stabilize soil.

3.3.4. Fly Ash Stabilization to Improve Soil Strength

Numerous projects have effectively used fly ash to enhance the soil's strength properties. Fly ash is useful for stabilizing embankments to increase slope stability, bases or subgrades to lessen lateral earth pressures, and backfill to stabilize slopes. The normal depth of stabilized soil is between 15 and 46 centimeters or 6 and 18 inches. Fly ash is primarily used in soil stabilization applications to increase the soil's compressive and shearing strengths. The compressive strength of fly ash-treated soils is dependent on (Figure 3.5):

- In-place soil properties
- Delay time
- Moisture content at time of compaction
- Fly ash addition ratio

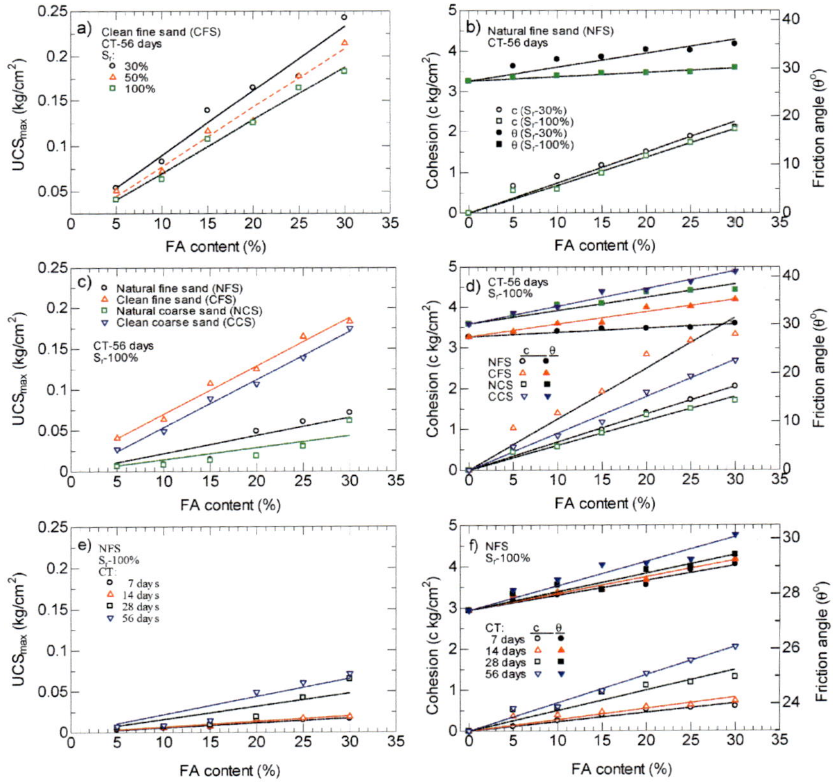

Figure 3.5. *The effect of fly ash percentages on the mechanical properties of fly – ash – stabilized sand.*

Note: (a) UCS_{max} – FA relationship of CFS at various S_r with CT – 56 days; (b) shear strength parameters – FA relationship of NFS at different S_r with CT – 56 days; (c) UCS_{max} – FA relationship of dissimilar sands with S_r – 100% and CT – 56 days; (d) shear strength parameters – FA relationship of diverse sands with

S_r – 100% and CT – 56 days, (e) UCS_{max} – FA relationship of NFS at various CT with S_r – 100%,; and (f) shear strength parameters – FA relationship of NFS at any kinds of CT with S_r – 100%.

Source: Simatupang M, Mangalla LK, Edwin RS, Putra AA, Azikin MT, Aswad NH, Mustika W. (2020). The mechanical properties of fly-ash-stabilized sands. Geosciences, 10(4):132. https://doi.org/10.3390/geosciences10040132.

3.3.4.1. Delay Time

The amount of time that passes between the fly ash and water mixture's initial contact and its ultimate compaction is known as the delay time. The compressive strength exhibits a strong correlation with delay time. The strength and density decrease as the final compaction date approaches. The tricalcium aluminate (C3A) reaction that happens quickly when Class C fly ash and water are combined makes delay time crucial.

Because some of the cementation potential is lost and some of the compactive energy must be used to overcome the soil particle bonding caused by cementation, densities and strengths are decreased. In soil-fly ash mixtures, maximum strength is immediately achieved. For construction purposes, a one-hour compaction delay is typically specified.

3.3.4.2. Moisture Content

The strength of the fly ash stabilized soil mixture is influenced by its water content. In soil-fly ash mixtures, the maximum strength is typically achieved at moisture contents lower than the ideal moisture content for density. The ideal moisture content for strength in silt and clay soils is typically four to 8% lower than the ideal moisture content for maximum density.

The ideal moisture content for granular soils is often 1 to 3% less than the ideal moisture for density. Controlling the moisture content during construction is therefore essential. Typically, a nuclear density measurement tool is used to determine the moisture content (Figure 3.6).

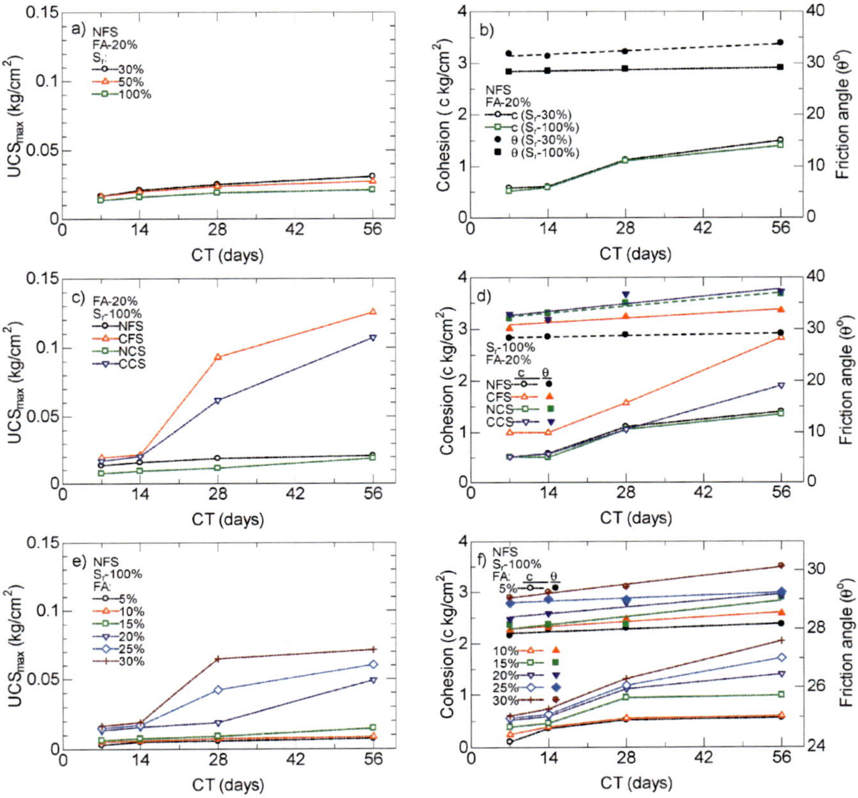

Figure 3.6. *The effect of curing time on the mechanical properties of fly – ash – stabilized sands.*

Note: (a) UCS_{max} – CT relationship of NFS at various S_r with 20% FA, (b) shear strength parameters – CT relationship of NFS at different S_r with FA 20%, (c) UCS_{max} – CT relationship of dissimilar sands with S_r – 100% and 20% FA, (d) shear strength parameters – CT relationship of diverse sands with S_r – 100% and 20% FA, (e) UCS_{max} – FA relationship of NFS at various FA with S_r – 100%, (f) shear strength parameters – FA relationship of NFS at any kinds of FA with S_r – 100%.

Source: Simatupang M, Mangalla LK, Edwin RS, Putra AA, Azikin MT, Aswad NH, Mustika W. (2020). The mechanical properties of fly-ash-stabilized sands. Geosciences, 10(4), 132. https://doi.org/10.3390/geosciences10040132.

3.3.4.3. Addition Ratios

Fly ash is typically added at rates between 8 and 16% depending on the dry weight of the soil. The desired strength, the fly ash's properties, and the type of soil all affect how much is added. Testing of the laboratory mix design is required to ascertain the addition rate. Generally speaking, the realized compressive strength increases with increasing addition rate. Although the requirements of this specification are not necessary for this application and could raise the cost of ash supply, fly ashes for State Department of Transportation projects are typically specified to meet AASHTO M 295 (ASTM C 618). More and more materials that do not comply with AASHTO M 295 are being employed with success. Note that almost any fly ash with some degree of self-cementitious properties can be designed to work well in transportation-related projects.

3.3.4.4. Soil Properties

The types of clay minerals present in the soil and the water that has been adsorbed by them affect the plasticity of soils treated with Class C or other high-calcium fly ash. In certain applications, soils with a sulfate content of more than 10% have been known to swell excessively. Furthermore, it is challenging to stabilize organic soils with fly ash.

3.3.5. Advantages of Fly Ash Stabilization

There are various benefits to stabilizing soil with fly ash. First of all, as fly ash is a byproduct of burning coal and is widely available in certain areas, it is an affordable solution. By reducing the need for more costly stabilizers like cement and lime, the use of fly ash lowers the overall cost of construction. Second, stabilizing fly ash has positive environmental effects. Fly ash's environmental impact is lessened because the process repurposes an industrial byproduct, which lessens the need to dump it in a landfill. Furthermore, expansive clays and loose sandy soils—two types of soils with poor engineering qualities—can perform better after being stabilized with fly ash, making them more suitable for building.

3.3.6. Applications of Fly Ash Stabilization

Stabilization of fly ash is widely used in building foundations, embankments, and roadways. Fly ash is frequently used in road construction to stabilize subgrade soils, which are the underlying natural soils that support the pavement layers. Fly ash stabilization helps to prevent pavement distress and extend the life of the road by strengthening and stabilizing the subgrade. Fly ash stabilization can lessen soil settling and increase its load-bearing capacity during embankment construction, making the soil more suited to supporting large, weighty structures. Additionally, soft and expansive soils—which are prone to swelling and shrinking with changes in moisture content—are stabilized with fly ash. Fly ash stabilization lessens these soils' plasticity and swelling potential, which helps shield structures built on these types of soils from harm.

3.3.7. Challenges and Limitations

Fly ash stabilization has many benefits, but it also has drawbacks and restrictions. The variation in fly ash's composition and quality is one of the main problems. Depending on the type of coal used and the combustion conditions, fly ash can have a very different chemical composition. Because of this variability, fly ash must first undergo extensive testing and characterization in order to determine its suitability as a stabilizer. The possibility for fly ash to leach trace elements into the groundwater and soil is another drawback. Although the majority of studies have demonstrated that there is little trace element leaching from fly ash-stabilized soils, this is still a concern that requires appropriate testing and monitoring (Doty and Alexander, 1968; Winterkorn, 1975; Zuber et al., 2013).

3.4. BITUMINOUS STABILIZATION

An inventive technique for improving the strength and durability of soil in a variety of engineering and construction projects is bituminous soil stabilization. This procedure entails enhancing the physical characteristics of the soil by incorporating bituminous materials, like asphalt. As a result, the soil base becomes more resilient, stable, and weather-resistant, making it perfect for a variety of uses, particularly in the building of roads and other infrastructure.

3.4.1. Benefits of Bituminous Soil Stabilization

1. **Increased Stability and Load Bearing Capacity:** The soil's ability to support large loads is greatly increased by bituminous stabilization, which qualifies it for heavy construction.
2. **Durability and Longevity:** The soil gains durability from the bituminous component, which increases its resistance to environmental elements like temperature changes and water erosion.
3. **Cost-Effectiveness:** Bituminous soil stabilization can be more affordable than other stabilization methods, particularly for large-scale projects where long-term durability is crucial.
4. **Environmental Considerations:** Compared to conventional methods, modern bituminous materials often include recycled components, making them more environmentally friendly.

3.4.2. Applications of Bituminous Soil Stabilization

1. **Road Construction:** Building strong road bases that can withstand heavy traffic and inclement weather is a common use for bituminous stabilization.
2. **Airport Runways:** The method provides a stable foundation for airport runways, ensuring a safe and durable surface for aircraft operations.
3. **Commercial and Residential Developments:** Bituminous stabilization can be used to prepare land for residential and commercial buildings in areas with poor soil quality.
4. **Agricultural and Rural Roads:** This technique is also beneficial for improving rural and agricultural roads, providing better connectivity and durability.

The process of stabilizing soil using bituminous materials is constantly evolving, providing economical and environmentally friendly solutions for soil stabilization challenges. Examples of these cutting-edge techniques and materials include bituminous emulsions and bituminous stabilized material for roads. Bituminous soil stabilization appears to have a bright future as technology develops, with potential uses extending into new areas of engineering and construction.

3.4.3. Advanced Techniques in Bituminous Soil Stabilization

Significant progress has been made in the field of bituminous soil stabilization, bringing with it more effective and environmentally friendly techniques. These cutting-edge methods solve many issues with conventional stabilization techniques while also improving the performance of stabilized soils.

1. **Utilizing Bituminous Emulsions:** Bituminous emulsions, which are composed of bitumen, water, and an emulsifying agent, are mixed with soil in this method. Emulsions ensure a more even distribution of the bituminous binder and improve the coating of soil particles.
2. **Foamed Bitumen Technology:** The process of producing foamed bitumen involves adding tiny amounts of air and water to hot bitumen, which causes it to expand. When combined with soil, this foamed bitumen provides exceptional stabilizing and bonding qualities, particularly for repurposing old road materials.
3. **Polymer Modified Bitumen (PMB):** Bitumen's elasticity, temperature stability, and resistance to water damage are all improved when PMB adds polymers to it. In regions where the weather is severe, this method is especially helpful.

4. **Geo-Synthetics and Bituminous Composites:** It has been shown to be very successful to combine bituminous materials with geo-synthetics like geotextiles and geogrids. This composite material keeps the soil from deforming or cracking and increases its tensile strength.
5. **Sustainable and Recycled Materials:** Reclaimed asphalt pavement (RAP) and other waste products are examples of recycled materials that are incorporated into bituminous soil stabilization, which is an environmentally friendly approach. This lowers the cost of materials and decreases waste.

The development of stronger and longer-lasting infrastructure has been made possible in large part by these cutting-edge methods. They play a particularly important role in large-scale land development and road construction projects, where the stability and durability of the foundation are critical. Furthermore, bituminous soil stabilization is a more environmentally friendly option in contemporary building practices because these techniques are in line with environmental concerns. The ongoing advancements in this area promise even more creative approaches to soil stabilization that take into account the effects on the environment as well as engineering difficulties.

3.4.4. Comparative Analysis of Bituminous Stabilization vs. Traditional Methods

When it comes to soil stabilization, bituminous stabilization and conventional techniques have unique benefits and drawbacks. This study explores many topics, including cost, effectiveness, environmental impact, and application suitability.

1. Effectiveness in Soil Stabilization:
   - **Bituminous Stabilization:** Provides excellent waterproofing and binding qualities. It improves the soil's ability to support loads and stabilizes a variety of soil types more successfully, particularly when building roads.
   - **Traditional Methods (e.g., Lime or Cement Stabilization):** These techniques, while useful for some soil types, might not offer bituminous stabilization's degree of flexibility and waterproofing. They work better with soils that need to have their pH adjusted or have chemicals stabilized.
2. Cost-Effectiveness and Efficiency:
   - **Bituminous Stabilization:** Long-term cost savings may be achieved because of its durability and low maintenance needs.

Even though the initial outlay is greater, there will be long-term cost savings.

- **Traditional Methods:** Generally less expensive at first, but over time, especially in areas exposed to harsh weather or large loads, they may require more expensive maintenance.

3. Environmental Impact:
   - **Bituminous Stabilization:** Recycled materials are often used in modern bituminous techniques, which lessens their negative environmental effects. The release of volatile organic compounds (VOCs) during application, however, continues to raise concerns.
   - **Traditional Methods:** Cement and lime production are energy-intensive and contribute to CO_2 emissions. However, these methods do not emit VOCs during application.
4. Application Suitability and Flexibility:
   - **Bituminous Stabilization:** Extremely appropriate for areas experiencing high traffic, airport runways, and road construction. It provides performance and application flexibility in a range of weather scenarios.
   - **Traditional Methods:** Ideal for small roads, building foundations, and locations requiring a high compressive strength. In very wet or dry conditions, their efficacy may be compromised.
5. Durability and Longevity:
   - **Bituminous Stabilization:** Provides enhanced durability and resistance to weathering, making it ideal for long-term infrastructure projects.
   - **Traditional Methods:** While durable, they may be susceptible to cracking and erosion over time, especially under fluctuating weather conditions.

3.4.5. Case Studies: Successful Bituminous Soil Stabilization Projects

Throughout the world, bituminous has been successfully applied in many cases to stabilize soil. Examining these case studies offers valuable perspectives on the applicability and efficacy of this approach in real-world scenarios.

1. Highway Construction in Tropical Regions:
   - Bituminous stabilization was used in a project in Southeast

Asia to overcome the difficulties of constructing roads in a humid, tropical environment. Road surfaces often deteriorate quickly due to the high moisture content of the soil.

- The road's durability was greatly increased and maintenance expenses were decreased by the use of bituminous materials, which created a layer that was resistant to moisture. This project proved bituminous stabilization's efficacy in addressing weather-related issues.

2. Airport Runway in Cold Climate:
   - Bituminous soil stabilization was used for a new runway at an airport in Northern Europe. The main issue was the extremely low temperatures in the area, which can cause soil to expand and contract, resulting in surface cracking.

   The runway demonstrated enhanced resistance to freeze-thaw cycles by incorporating bituminous materials into the soil, guaranteeing a more secure and dependable surface for aviation operations.

3. Commercial Development on Weak Soil:
   - Engineers in a North American commercial development project had to deal with the difficulty of building on unstable, compressible soil. Using traditional methods would have needed replacing a large amount of soil at a high cost.
   - The weak soil was turned into a stable foundation by applying bituminous soil stabilization, which also allowed large structures to be built without the need for new soil, saving a significant amount of time and money.

4. Rural Road Improvement in Developing Countries:
   - Bituminous soil stabilization has been used to improve rural road networks in many developing countries. Improving accessibility and connectedness in remote locations was the goal.
   - This approach provided a cost-effective and durable solution, improving the quality of life for rural communities by facilitating better access to markets, healthcare, and education.

These case studies highlight bituminous' adaptability and effectiveness for stabilizing soil under various climatic and geographic circumstances. They draw attention to how this approach not only satisfies a project's short-term requirements but also advances cost-effectiveness and sustainability over the long run. Because of this, bituminous soil stabilization is still the material of choice for many infrastructure and building projects across the globe.

3.5. POLYMER STABILIZATION

Long-chain molecules called polymers are commonly employed in soil stabilization. These molecules can form a network with soil particles, improving the cohesion and bonding within the soil matrix. Because of this interaction, polymer-stabilized soils have better mechanical qualities, such as greater strength and decreased plasticity, which increase their resistance to deformation under load. Due to the adaptability of polymers, they can be applied in a variety of soil types, including clay and sandy soils, which makes this technique appropriate for a range of civil engineering projects.

3.5.1. Properties of Polymers Used in Stabilization

The kind of polymer utilized has a significant impact on how well polymer stabilization works. Polymers fall into a number of categories: biopolymers, synthetic organic polymers (like polyacrylamide and polyethylene), and natural polymers (like cellulose and starch). Every kind of polymer has distinct qualities that can affect how well it stabilizes soil.

1. **Synthetic Organic Polymers:** Because of their great strength, longevity, and ease of application, these polymers are the most often used ones in soil stabilization. For instance, polyacrylamide is frequently utilized due to its remarkable capacity to lower soil erosion and increase water retention. Polyethylene is another frequently used synthetic polymer that has good chemical and water resistance, which makes it perfect for stabilizing soils in abrasive environments.
2. **Biopolymers:** Biopolymers, which are synthetic polymer substitutes made from natural sources such as microorganisms and plants, are becoming more and more popular. Guar gum, chitosan, and xanthan gum are a few examples. These biopolymers are appropriate for projects where sustainability is a top concern because they are biodegradable and have a minimal environmental impact. Although biopolymers might not offer the same strength as synthetic polymers, they work well in applications that call for a moderate amount of stabilization.
3. **Natural Polymers:** Natural polymers such as cellulose and starch have been used for centuries to stabilize soil, though they are less common in modern engineering. These materials are plentiful and biodegradable, but mechanical performance is usually not as good as it can be when using synthetic materials. To improve their stabilizing qualities, natural polymers are being mixed and altered in response to developments in polymer technology.

3.5.2. Mechanisms of Polymer Stabilization

Polymer stabilization works through several mechanisms that enhance the physical and mechanical properties of soil. These mechanisms include adhesion, encapsulation, flocculation, and film formation.

1. **Adhesion:** Soil particles can stick to polymers, which binds the particles together to form a cohesive matrix. This strengthens the soil's shear strength and decreases its plasticity, increasing its resistance to deformation under load. When it comes to stabilizing loose or granular soils, the adhesive qualities of polymers are especially helpful because they help to stop erosion and particle displacement.
2. **Encapsulation:** Polymers have the ability to enclose soil particles in certain situations, forming a barrier that protects the soil from external elements like moisture and temperature changes. This encapsulation effect helps to preserve the integrity of the soil structure, which makes it particularly helpful for stabilizing soils in areas with heavy rainfall or extreme temperatures.
3. **Flocculation:** The process by which polymers induce fine soil particles to coalesce into larger aggregates is known as flocculation. As a result, the soil has fewer fine particles, which may both increase and decrease the soil's permeability and plasticity. Clayey soils, which are prone to swelling and shrinking because of their high plasticity, benefit greatly from flocculation.
4. **Film Formation:** Certain polymers can reduce the permeability of the soil and increase its resistance to erosion by forming a thin film over the soil particles. In addition to controlling the water content of the soil and averting problems like frost heave and soil swelling, this film serves as a barrier to moisture.

3.5.3. Advantages of Polymer Stabilization

Polymer stabilization offers several advantages over traditional soil stabilization methods, making it an attractive option for a wide range of civil engineering applications.

1. **Environmental Sustainability:** The environmental sustainability of polymer stabilization is one of its biggest benefits. Many polymers can be produced with lower energy inputs and have a smaller carbon footprint than cement and lime, which require high-energy production processes and contribute to carbon dioxide emissions. Furthermore, because biopolymers are biodegradable and made from

renewable resources, their use further improves the sustainability of this process.

2. **Improved Soil Properties:** Strength, cohesiveness, and erosion resistance are just a few of the mechanical qualities of soil that can be markedly enhanced by polymer stabilization. Because of this, polymer-stabilized soils are stronger, more suitable for sustaining structures, and more resistant to environmental stresses. Because polymers are flexible, it is also possible to fine-tune the properties of the soil to meet the requirements of a particular project, something that is not always possible with traditional stabilizers.
3. **Versatility:** Any kind of soil, from expansive clays to sandy soils, can be stabilized with polymers. Because of its adaptability, polymer stabilization is a good choice for a range of civil engineering tasks, such as erosion control, slope stabilization, and road building. Moreover, polymers can be applied in a variety of ways, including mixing, injecting, and spraying, making them adaptable to a range of site conditions.
4. **Reduced Construction Time:** When compared to conventional methods, polymer stabilization can be applied quickly and effectively, cutting down on construction time. This is especially helpful for projects with short timelines or when quick stabilization is needed. Certain polymers have quick setting times, which facilitate quicker advancement to later stages of construction.
5. **Water Resistance:** Many polymers have exceptional water resistance, which makes them perfect for use in regions that frequently experience flooding or heavy rains. Polymers aid in preventing water infiltration, which can cause problems like frost heave, soil swelling, and loss of bearing capacity. They do this by lessening the permeability of the soil.
6. **Cost-Effectiveness:** Polymer stabilization can be a more affordable option in the long run due to its advantages over traditional stabilizers, such as lower maintenance costs and a longer stabilized soil lifespan. However, the initial cost of polymers may still be higher. Furthermore, overall savings may arise from the use of polymers in smaller amounts and at a lower cost of application.

3.5.4. Applications of Polymer Stabilization

Polymer stabilization is used in a variety of civil engineering applications, particularly where traditional stabilization methods may be less effective or where environmental sustainability is a priority.

1. **Road and Pavement Construction:** One of the most common applications of polymer stabilization is in road and pavement construction. Polymers are used to stabilize the subgrade and base layers, improving the load-bearing capacity and durability of the pavement. The water resistance provided by polymers also helps to extend the lifespan of roads by preventing water infiltration and the associated issues of frost heave and soil erosion.
2. **Slope Stabilization:** Polymers are often used in slope stabilization projects to prevent soil erosion and landslides. The adhesive and cohesive properties of polymers help to bind soil particles together, reducing the risk of slope failure. Additionally, the encapsulation and film formation effects of polymers protect the soil from moisture and temperature changes, further enhancing slope stability.
3. **Erosion Control:** Polymer stabilization is an effective method for controlling soil erosion, particularly in areas prone to heavy rainfall or wind erosion. Polymers can be applied to the soil surface to form a protective layer that reduces erosion by wind and water. This application is commonly used in agriculture, construction sites, and along riverbanks and shorelines.
4. **Dust Suppression:** In arid regions or construction sites, polymer stabilization can be used for dust suppression. Polymers are sprayed onto the soil surface, forming a crust that prevents dust particles from becoming airborne. This not only improves air quality but also helps to protect the surrounding environment and reduce the health risks associated with dust inhalation.
5. **Foundation Stabilization:** Polymer stabilization can be used to improve the bearing capacity of soils used in foundation construction. By enhancing the strength and cohesion of the soil, polymers help to prevent settlement and increase the load-bearing capacity of the foundation, making them suitable for use in both residential and commercial construction.
6. **Landfill Liners and Covers:** Polymers are also used in the construction of landfill liners and covers, where their low permeability and resistance to chemical attack are essential for preventing leachate contamination and protecting the environment. Polymer-stabilized soils are often used as a barrier layer in landfill covers, helping to reduce the risk of groundwater contamination.

3.5.5. Challenges and Limitations

Polymer stabilization has many benefits, but it also has drawbacks and restrictions. Variability in soil conditions can impact polymer performance, making it one of the main challenges. It's vital to carefully choose the right polymer for each project because different soil types may call for different kinds of polymers or application techniques. The potential for environmental impact is another problem, especially when it comes to synthetic polymers. Even though a lot of synthetic polymers are strong and long-lasting, their lack of biodegradability raises potential long-term environmental issues. Although using biopolymers can help with this problem, their performance might not always be on par with that of synthetic alternatives.

Furthermore, polymer stabilization may initially cost more than more conventional techniques, especially when high-performance polymers are being used or when large amounts of material are needed for a project. However, these upfront expenses can be compensated for by the long-term advantages and possible cost savings from less maintenance and a longer lifespan (Doty and Alexander, 1968; Winterkorn, 1975; Zuber et al., 2013).

REFERENCES

1. Balasubramaniam, S., Buensuceso, B. R., Phien-Wej, N., & Bergado, D. T. (1990). Engineering behavior of lime-stabilized soft Bangkok clay. In *Proceedings of the 10th Southeast Asian Geotechnical Conference* (Vol. 1, pp. 23–28).
2. Basma, A. A., & Tuncer, E. R. (1991). Effect of lime on volume change and compressibility of expansive clays. *Transportation Research Record, 1295.*
3. Bell, F. G. (1989). Lime stabilization of clay soils. *Bulletin of the International Association of Engineering Geology, 39*(1), 67–74.
4. Chen, L., & Lin, D. F. (2009). Stabilization treatment of soft subgrade soil by sewage sludge ash and cement. *Journal of Hazardous Materials, 162*(1), 321–327.
5. Chiu, K. H., & Chin, K. Y. (1963). The study of improvement bearing capacity of Taipei silt by using quicklime piles. In *Proceedings of the 2nd Asian Regional Conference on Soil Mechanics and Foundation Engineering* (Vol. 1, pp. 387–393).
6. Croft, J. B. (1964). The processes involved in the lime stabilization of clay soils. In *Australian Road Research Board (ARRB) Conference* (Vol. 2, No. 2). Melbourne.
7. Dawson, R. F., & McDowell, C. (1961). A study of an old lime-stabilized gravel base. In *Lime Stabilization: Properties, Mix Design, Construction Practices and Performance* (Bulletin 304). Highway Research Board.
8. Dawson, R. F., & McDowell, C. (1961). A study of an old lime-stabilized gravel base. *Highway Research Board Bulletin, 304.*
9. Doty, R., & Alexander, M. L. (1968). Determination of strength equivalency for design of lime-stabilized roadways. *Report No. FHWA-CATL-78-37.*
10. Doty, R., & Alexander, M. L. (1968). Determination of strength equivalency for the design of lime-stabilized roadways. *Report No. FHWA-CA-TL-78-37.*
11. Eades, J. L., & Grim, R. E. (1960). Reaction of hydrated lime with pure clay minerals in soil stabilization. *Highway Research Board Bulletin, 262.*
12. Eades, J. L., & Grim, R. E. (1996). A quick test to determine lime requirements for lime stabilization. *Highway Research Board, National Research Council,* (139), 61–72.
13. Eisazadeh, H., Kassim, K. A., & Nur, H. (2012). Solid-state NMR and FTIR studies of lime-stabilized montmorillonitic and lateritic clays. *Applied Clay Science, 67*, 5–10.

14. George, S. Z., Ponniah, D. A., & Little, J. A. (1992). Effect of temperature on lime-soil stabilization. *Construction and Building Materials, 6*(4), 247–252.

15. Graves, R. E., Eades, J. L., & Smith, L. L. (1988). Strength developed from carbonate cementation of silica-carbonate base course materials. *Transportation Research Record, 1190.*

16. Herrier, G., Lesueur, D., Puiatti, D., Auriol, J. C., Chevalier, C., Haghighi, I., & Fry, J. J. (2012). Lime-treated materials for embankment and hardfill dam. In *ICOLD 2012–International Symposium on Dams for a Changing World*, Kyoto (pp. 5–8).

17. Herrin, M., & Mitchell, H. (1961). Lime-soil mixtures. *Highway Research Board Bulletin, 304*, 99–138.

18. Holtz, W. G. (1969). Volume change in expansive clay soils and control by lime treatment. In *Proceedings of the 2nd International Research and Engineering Conference on Expansive Clay Soils* (pp. 157–174).

19. Imran, M. S., Gary, K. F., & Michael, P. E. (2007). Innovation in cement stabilization of airfield subgrades. In *Proceedings of FAA Worldwide Airport Technology Transfer Conference* (pp. 6–8). Atlantic City, NJ.

20. Little, D. N. (1993). Evaluation of the structural properties of stabilized pavement layers. *Interim Report to the Texas Department of Transportation Research Project 1287.*

21. Mallela, J., Quintus, H. V., & Smith, K. (2004). Consideration of lime-stabilized layers in mechanistic-empirical pavement design. *The National Lime Association.*

22. Mateos, M. (1964). Soil-lime research at Iowa State University. *Journal of the Soil Mechanics and Foundations Division, 90*(2), 127–156.

23. Maxwell, S. (2005). Geosynthetics in stabilizing soft subgrade with breaker run. *Wisconsin Highway Research Program, Final Report No. 0092-45-15*, 88.

24. McCallister, L. D., Knodel, P. C., & Petry, T. M. (1992). Leach tests on lime-treated clays. *Geotechnical Testing Journal, 15*(2), 106–114.

25. Nalbantoglu, Z., & Tuncer, E. R. (2001). Compressibility and hydraulic conductivity of a chemically treated expansive clay. *Canadian Geotechnical Journal, 38*(1), 154–160.

26. Nowlin, L., Smith, R. E., & Little, D. N. (1992). Back-calculated in-situ moduli of lime-stabilized layers from FWD data in Texas pavements. *Texas Transportation Institute.*

27. Parkash, E. V., & Kumar, E. V. (2017). Use of lime and sawdust ash in

soil stabilization. *International Journal of Innovative Research in Science, Engineering and Technology, 6*(2), 1682–1689.

28. Rajasekaran, G., & Narasimha Rao, S. (2000). Strength characteristics of lime-treated marine clay. *Proceedings of the Institution of Civil Engineers–Ground Improvement, 4*(3), 127–136.
29. Sadek, D., Roslan, H., & Abubakar, D. A. (2008). Engineering properties of stabilized tropical peat soils. *Electronic Journal of Geotechnical Engineering, 13*, 7–8.
30. Sai Kumar, V. B. Y., & Hanumantharao, C. (2017). Behavior of building frames on soils of varying plasticity index. *International Journal of Civil Engineering and Technology, 8*(1), 623–627.
31. Saitoh, S. Y., Suzuki, & Shirai, K. (1985). Hardening of soil improvement by deep mixing method. In *Proceedings of the 11th International Conference on Soil Mechanics and Foundation Engineering* (Vol. 3, pp. 1745–1748). San Francisco, CA.
32. Seed, H. B., Woodward, R. J., & Lundgren, R. (1978). Prediction of swelling potential for compacted clays. *Transactions of the American Society of Civil Engineers, 128*(1), 1443–1477.
33. Sivapullaiah, P. V., Prashanth, J. P., & Sridharan, A. (1996). Effect of fly ash on the index properties of black cotton soil. *Soils and Foundations, 36*(1), 97–103.
34. Syed Zuber, S. Z., Kamarudin, H., & Abdullah, M. M. (2013). Review on soil stabilization techniques. *Australian Journal of Basic and Applied Sciences, 7*(5), 258–265.
35. Terashi, M., Tanaka, H., & Kitazume, M. (1983). Extrusion failure of ground improved by the deep mixing method. In *Proceedings of the 7th Asian Regional Conference on Soil Mechanics and Foundation Engineering* (Vol. 1, pp. 313–318). Haifa, Israel.
36. Thompson, M. R. (1968). Lime-treated soils for pavement construction. *Journal of the Highway Division, 94*(2), 191–218.
37. Tuncer, E. R., & Basma, A. A. (1991). Strength and stress-strain characteristics of a lime-treated cohesive soil. *Transportation Research Record, 1295*.
38. Wild, S., Kinuthia, J. M., Jones, G. I., & Higgins, D. D. (1999). Suppression of swelling associated with ettringite formation in lime-stabilized sulfate-bearing clay soils by partial substitution of lime with ground granulated blast furnace slag (GGBS). *Engineering Geology, 51*(4), 257–277.
39. Winterkorn, H. F. (1975). Soil stabilization. In H. F. Winterkorn & F.

Y. Fang (Eds.), *Foundation Engineering Handbook* (pp. 312–336). Van Nostrand Reinhold.

40. Zolkov, E. (1962). Influence of chlorides and hydroxides of calcium and sodium on consistency limits of a fat clay. *Highway Research Board Bulletin, 309.*

CHAPTER 4

MECHANICAL STABILIZATION METHODS

Contents

INTRODUCTION

Mechanical stabilization methods in road construction are crucial techniques used to enhance the performance and durability of roadways. These methods focus on improving the strength, stability, and longevity of road surfaces and subgrades, which are vital for withstanding the stresses imposed by traffic loads, environmental conditions, and soil characteristics.

One primary mechanical stabilization technique is soil compaction. This involves using heavy machinery, such as rollers or compactors, to increase the density of the soil or aggregate materials beneath the road surface. Proper compaction reduces soil compressibility, minimizes settlement, and enhances load-bearing capacity, which is essential for supporting the road structure and preventing issues such as rutting and potholes.

Another key method is the use of stabilizing additives. These additives, such as lime, cement, or fly ash, are mixed with soil or aggregate materials to improve their engineering properties. Lime stabilization, for instance, involves mixing lime with soil to enhance its strength and reduce its plasticity, which improves the soil's ability to support traffic loads and resist moisture-induced deformation.

Geosynthetics, including geotextiles and geogrids, are also employed in road construction for mechanical stabilization. Geotextiles are synthetic fabrics placed between soil layers to provide separation, reinforcement, and filtration, while geogrids are used to stabilize soil by distributing loads and preventing deformation. These materials enhance the structural integrity of the roadbed, reduce erosion, and extend the lifespan of the road.

Mechanical stabilization methods in road construction are vital for ensuring road durability and performance. By employing techniques such as soil compaction, stabilizing additives, and geosynthetics, engineers can effectively address challenges related to soil stability, load-bearing capacity, and environmental conditions, ultimately leading to more reliable and long-lasting roadways.

4.1. MECHANICAL SOIL STABILIZATION

The physical process of mechanical stabilization modifies the physical characteristics of the soil structure. Additives are used in chemical stabilization to change the chemical properties of the soil, making it more stable.

Compaction is a technique used in mechanical soil stabilization techniques to interlock soil-aggregate particles. When soil is compacted, a dense mass must be created by the distribution of soil particle sizes. Compaction after homogeneous mixing will yield stabilized soil. A dense, well-graded soil-aggregate mixture can be created by adding more fines or aggregates. Geotextiles can occasionally be used to enhance the soil's engineering properties.

The process of improving an existing material by mixing it with one or more other materials to enhance its plasticity and particle size distribution is known as mechanical stabilization. River-deposited sand, natural gravel, silty sands, sand clays, waste quarry products, crushed run quarry products, scoria, volcanic cinders, poorly graded laterites, beach sands, etc., are typical materials used for mechanical stabilization. Blended materials are evaluable using standard techniques and have properties comparable to those of conventional unbounded materials.

Internal friction and cohesiveness are the main characteristics influencing the stability of compacted base or sub-base materials. Primary factors influencing internal friction are the features of the larger soil particles, i.e., sizes of silt, sand, and gravel. As shown by the plastic properties, the cohesion, shrinkage, swelling, and compressibility are primarily related to the amount and type of clay fraction.

Particle size distribution and plastic properties form the basis of the initial mix design for mechanical stabilization. Additionally, it is preferred that strength tests (CBR, etc.) be conducted to confirm that the necessary improvement has been made. The use of unconventional materials frequently necessitates more thorough testing and investigation, which may involve changing the approved design or specification requirements.

4.1.1. Particle Size Distribution

While achieving a high density will typically result in a high frictional strength, maximum density and maximum frictional strength are not always the same. It has been demonstrated experimentally that a particle size distribution that yields the maximum dry density (MDD)—achieved with the closest packing and the fewest voids—follows Fuller's equation, with the value of the exponent "n" typically falling between 0 and 50 for the majority of soils. But with certain materials, like, e.g., high densities can be attained with "n" values as low as 0.33 in the gravel-sand-clays mixture. The percentage of fines that pass through the 75 μm sieve for materials with a maximum size of 19 mm is 6 and 8% for "n" values of 0.5 and 0.45, respectively. Higher percentages of material passing the 75 μm sieve might work best in some circumstances.

The materials utilized should have a particle size distribution within the bounds defined by replacing the value of "n" in the above equation with the values of "0.50" and "0.33" when the pavement design depends on a relatively low permeability in the pavement courses. These bounds are wide enough to accommodate the inevitable variations in field mixing. However, stricter restrictions might be imposed if plant mixing is done. The material's fines content may be excessive if the exponent "n" value is less than 0.33. Reduced permeability will be the outcome of high fine content, but it may also cause pore pressures to develop and instability during compaction or in service. When "n" is larger than 0.5, the material is typically harsher, more likely to segregate and ravel, and more challenging to work with.

4.1.2. Liquid Limit and Plasticity Index

For mechanically stabilized materials, the plasticity limits listed for the various pavement structure layers can typically be used as satisfactory design criteria. Furthermore, there is typically a relationship between the Plasticity Index and Linear Shrinkage of a material passing a 0.425 mm sieve. The allowable shrinkage values can be estimated from the allowable PI values or found through testing. For sealed pavements, the typical values are 2%, and for unsealed pavements, 3%.

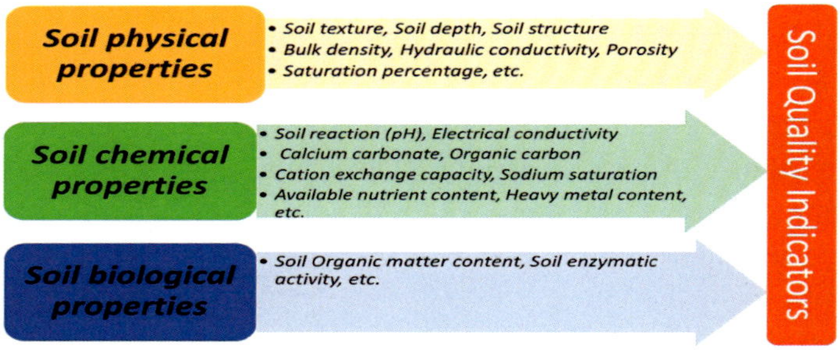

Figure 4.1. *General analysis.*

Source: El Behairy RA, El Arwash HM, El Baroudy AA, Ibrahim MM, Mohamed ES, Kucher DE, Shokr MS. (2024). How can soil quality be accurately and quickly studied? A review. Agronomy, 14(8):1682. https://doi.org/10.3390/agronomy14081682.

For gravel to receive bituminous surface treatment when the percentage of soil binder is low, the Plasticity Modulus (PI x the percentage passing the 0.425 mm sieve relative to the whole material) should, as a general rule, not exceed 200. If road formations are well-drained, it may be possible to relax the PM to roughly 400 in arid climates. If the soil binder possesses certain natural setting properties, then slightly broader limits of the Plasticity Index might work well with certain ironstone gravels and limestone rubbles. This should not be interpreted as a general rule, though; rather, each case should be considered on its own merits, and care should be taken when working with novel or unfamiliar materials. When undertaking major construction projects, it is best to build trial pavement sections for assessment at least two years before starting their extensive use (Figure 4.1).

4.1.3. Strength Tests

Stabilized materials can be evaluated using appropriate strength tests at the density and moisture content of the pavement at the end of its service life. Predicting the moisture condition of a potential base material is crucial when testing it. This method allows for the derivation of the failure envelopes at moisture conditions that border on equilibrium moisture conditions and at the necessary density expected in the suggested pavement. It is possible to determine the equilibrium moisture conditions that should be present in a pavement by looking at current roads made of materials that are comparable to those under investigation and compiling this data for later use.

The laboratory CBR test is one of the most widely used strength tests. It has been discovered that the values listed in Table 4.1 are generally applicable. Typically, compacted specimens are soaked for four days before testing. To mimic the worst circumstances anticipated in service, the test's conditions may be changed in terms of moisture content and degree of compaction.

For relatively impermeable materials, soaking the specimens for four days may be too conservative in some situations, while longer times may be more appropriate. In this situation, it would be wise to adopt a minimum CBR value that is different from the ones shown in the Table 4.1.

Table 4.1. *California Bearing RATIO Limits for Mechanically Stabilized Base Materials*

Pavement	Minimum CBR values
High class, high traffic volume	100
Rural roads, wet areas	80
Rural roads, dry areas	60

Source. https://wollocivil.wordpress.com/wp-content/uploads/2014/04/chapter-6-stabilized-pavement-materials.pdf.

Local experience, particularly that about the performance of local materials, should be considered when choosing appropriate criteria. Characterizing each material separately, adjusting their proportions to meet the chosen parameters, and creating a trial mixture to ensure that the ideal ratios actually yield the required properties are all part of the design process for stabilized mixtures. If a satisfactory road pavement is to be produced, then proper construction and control testing techniques are just as important as sufficient investigation and design. The process entails meticulously weighing and blending the component parts to create a homogeneous, unsegregated end product that can be compressed and completed in compliance with the specification (De Groot et al., 1996; Han and Collin, 2005).

4.2. SOIL COMPACTION TECHNIQUES

Because soil compaction is such a skill, even seasoned operators can make mistakes. The secret is to cover a region with enough passes to achieve the required density without going overboard. That is a time waster. The proper amount of soil must be added to the trench backfill, and the appropriate amount of force must then be applied. Soil that is overexposed will not compact evenly. The best outcomes will come from carefully planning each soil compaction project. Know the type of soil you are working with. Select the appropriate tool for the work. And pick a knowledgeable operator. In this manner, you can be sure that the road construction project is off to the best possible start.

Road construction season has returned. The procedure appears uncomplicated to individuals who are not connected to the construction sector. Within, though, knows better. Things can go horribly wrong later on, so to speak if the foundation of the roadway is not done correctly. Proper compaction of the soil is essential.

Tamping down the underlying soil has the purpose of increasing its density by eliminating air pockets. A heavier load can be supported by dense, uniform soil without the risk of it shifting under the weight. Compaction also lessens water seepage, which can lead to issues such as winter frost damage. It is understandable why load-bearing capacity is important when one considers the weight of the automobiles and commercial trucks that frequently use our roads.

An inadequately prepared underlayment begs for issues. Concrete or asphalt can eventually buckle, bulge, or crack due to lack of maintenance. Because of structural flaws, the road will need to be replaced or repaired, which will be very expensive. Compaction of the soil properly preserves the integrity of the completed road.

Contractors have to consider several factors to ensure compaction goes well.

1. Moisture Content

The amount of moisture in different soils varies. The soil cannot be overly wet or dry to achieve the proper compaction. Soil, whether it's wet or dry, lacks stability to yield desired outcomes. There are three basic soil types:

- Cohesive, or clay
- Granular, or sand
- Organic

Organic soil is meant for gardening; it is not suitable for constructing roads. It is possible to compact granular and cohesive soils. Sand-filled soils have a tendency to crumble, while cohesive particles, as the name suggests, tend to stick together. At the job site, you can easily distinguish between the two by taking a handful of soil, balling it up in your fist, and letting it drop to the ground.

If it's overly dry, it will crumble as it falls and appear powdery even after you've squeezed it. When it hits the ground, it will stay in its balled-up shape if it is overly wet. Your hand might get muddy or wet from it. When you squeeze it, good compaction soil will be sufficiently damp to take on a shape, but it will break into a few pieces as it hits the ground.

The right amount of moisture in the soil allows for optimal compaction and expedites the completion of the task.

2. Choosing and Maintaining Compaction Equipment

There are several types of equipment you can use to compact soil:

- Rammers and trench rollers work best on cohesive soils
- Walk-behind rollers, ride-on rollers and forward/reversible plates work best on granular soils.

Whatever kind of equipment you use, upkeep is essential. Equipment malfunctions have the potential to stop the entire road project because soil compaction needs to be completed before moving forward. These devices put in a lot of effort. The weight of small compaction equipment can account for up to 20% of the force applied. Stated differently, a 185-lb. An area 11 feet square is compacted by a rammer, which exerts roughly 3,000 pounds of force.

Adhere to the manufacturer's recommended maintenance schedule, which usually calls for an annual service and inspection. If you rent compaction equipment, find out how frequently the dealership or rental shop maintains the machinery.

To improve the density, strength, and load-bearing capacity of the soil, a crucial step in the construction of roads is soil compaction. Roads with proper compaction are stable and long-lasting, resistant to rutting, cracking, and settlement in the future. To achieve the best possible road performance, understanding the theory underlying soil compaction as well as the practical methods applied in the field are crucial (De Groot et al., 1996; Han and Collin, 2005).

4.2.1. Theory of Soil Compaction

Soil acts as the subgrade or foundation for building roads, and to improve its structural qualities, it must be compacted. By decreasing the void spaces between soil particles, the process enhances load distribution and lowers the permeability of the soil. The ratio of the mass of solids to the volume they occupy, or the dry density of the soil, is commonly used to determine the degree of compaction.

Soil can be broadly categorized into two types for compaction purposes:

1. **Cohesive Soils (Clays and Silts):** These soils have small particle sizes and tend to hold moisture, making them more challenging to compact. They require careful control of moisture content to achieve optimal compaction.
2. **Granular Soils (Sands and Gravels):** These soils have larger particle sizes and are easier to compact. They can achieve good compaction with less moisture and are less prone to excessive water retention.

The ideal moisture content (OMC) for compaction is frequently found using the Proctor Test, both standard and modified. To ensure that the soil is neither too dry (which would result in poor compaction) nor too wet (which would cause instability), the test helps determine the moisture level at which soil can achieve MDD.

4.2.2. Factors Affecting Soil Compaction

- **Moisture Content**: The amount of water in the soil affects compaction efficiency. Too much water creates a slurry, while too little leads to insufficient cohesion.
- **Soil Type**: Granular soils compact more easily than cohesive soils due to particle size and shape.
- **Compaction Energy**: The amount of force applied, either through weight or vibration, impacts the level of compaction.

4.2.3. Practical Soil Compaction Techniques

1. **Static Compaction:** To compress the soil, static compaction requires

large machinery, like rollers or compactors, to operate continuously. When removing air voids is the main objective, the process works well for compacting coarse-grained soils. This technique often makes use of tools like pneumatic and smooth-wheel rollers.

- **Smooth-Wheel Rollers**: Used for compacting granular soils or subgrade layers, smooth-wheel rollers are effective in compressing soil without significant vibration.
- **Pneumatic Rollers**: These rollers have rubber tires and are useful for compacting fine-grained soils. They apply both pressure and kneading action, ensuring better compaction for cohesive soils.

2. **Dynamic Compaction:** Vibratory or impact energy is used in dynamic compaction to compact the soil. It works especially well for soil layers that are deeper and is commonly applied to both cohesive and granular soils. Tamping rollers, vibratory plate compactors, and vibratory rollers are examples of essential equipment.
   - **Vibratory Rollers**: These machines are equipped with a mechanism that generates vertical vibrations, allowing them to penetrate deeper layers of soil. The combination of vibration and weight is particularly effective for granular soils.
   - **Vibratory Plate Compactors**: These are smaller, hand-held or walk-behind machines used for compacting confined areas, such as trenches or around structures, where large rollers are impractical.
   - **Tamping Rollers**: These rollers are fitted with protruding feet or pads that compact the soil through a kneading action, ideal for cohesive soils.
3. **Impact Compaction:** To compact the soil, impact compaction entails lowering large weights from a height. This method is frequently applied in dynamic compaction, a process that transfers energy into the soil's deeper layers by repeatedly dropping heavy, dense objects like steel weights onto the ground. For large-scale soil densification, this technique works well for loose or weak soils.
4. **Kneading Compaction:** This technique works especially well with cohesive soils, where air voids must be removed by kneading or shearing. Kneading compaction is frequently applied using tools like sheepsfoot rollers, which press into the ground with their extended feet. Because clays and silts are cohesive materials that resist compression, the shearing motion aids in their compacting.

5. **Rammers and Hand-held Compactors:** Rammers and handheld compactors are used in places like trenches and small spaces where big machinery cannot work. These are compact but effective tools that compact soil in small spaces by using direct force. When building roads, they are especially helpful for compacting the earth around pipes, foundations, and other structures.

4.2.4. Monitoring and Quality Control

Thorough observation is necessary to guarantee that the soil has attained the appropriate stability and density for effective compaction. The dry density of the soil is typically measured *in-situ* using field density tests, such as the nuclear density test and the sand cone test. These tests verify that the soil has reached the ideal moisture content and maximum density and that the compaction satisfies the design requirements (De Groot et al., 1996; Han and Collin, 2005).

4.2.5. Methods of Compaction

Embankments were allowed to naturally compact due to a lack of attention to embankment design and construction. Heavy and quick compaction now requires the use of a high degree of sub-grade support, as the older models could not support heavier axle loads. Appropriate compacting equipment is required to carry out this type of compaction. Compaction, also known as densification, improves the properties of soil overall and enhances its performance as a pavement support bed. By mechanically compacting the soil particles closer together and minimizing air spaces, this process raises the density of the soil. By packing down the air spaces, a homogeneous soil mass with superior qualities is produced.

There are different methods and machinery used in the process of compaction.

4.2.5.1. Smooth Wheeled Roller

Typically, a smooth-wheel roller has three wheels: two large ones in the back and one small one up front. An internal combustion engine powers this device. The tandem smooth wheel roller, also known as the smooth wheel roller, is an additional variant of this device. Two drums are present in this version, one at the front and one at the back.

The smooth-wheeled rollers provide a smooth surface that can swiftly drain rainwater and are ideal for sealing fill surfaces at the end of the day. Deep soil layers, such as silty or uniform sands, are not suited for these machines. Their low compaction pressures make them unsuitable for deep layers. In deep layers, they will also result in stratification due to the uneven compaction. The smooth wheeled roller works best for compacting fine-grained, non-plastic soils and coarse, well-graded soils used in highway base courses.

4.2.5.2. Vibratory Rollers

A drum holds the vibratory rollers in place. There are two types of rollers available: smooth wheel and pneumatic tire. With the smooth wheel type, the drum oscillates at a high frequency and low amplitude through the use of a separate motor that drives a variety of eccentric weights. The vibratory roller with smooth wheels works well for compacting granular soils without layers up to one meter thick. The thickness will need to be decreased if the fines in the soil are noticeable.

- The wheel axle of the pneumatic roller is connected to a separate vibrating unit. The ballast box is not going to vibrate because it is suspended independently of the axle. Pneumatic-tired vibratory rollers are useful for compacting granular soils up to 30 centimeters in thickness. Granular soil can be compacted using this type of roller to a very high MDD.

4.2.5.3. Sheepsfoot Roller

The name "Sheepsfoot Roller" comes from a historical practice of compacting newly formed fill with a herd of sheep. This kind of roller resembles the smooth wheel roller quite a bit. The Sheepsfoot Roller differs from the other one in that it has multiple feet that protrude from 200 to 250 mm in length, as well as an end section that measures 40 to 65 cm. This roller has hollow drum wheels that can be filled with water and ballasts to increase their weight. The steel frame is where the drums are mounted.

The Sheepsfoot Roller applies pressure in the range of 7 to 42 kg/cm. Both plastic and nonplastic, fine-grained soils as well as coarse-grained soils that contain more than 20% fine particles work well with the Sheepsfoot Roller.

4.2.5.4. Impact Rammers

The ground is compressed by the impact rammer's impact force. The rammer will elevate its foot and then force it back to the ground. Deeper compaction depths produced by this force are perfect for confined areas like trenches.

Impact rammers are typically utilized in difficult-to-reach areas where other compactors might have trouble because of their unique design. They have little trouble moving around and covering the ground.

4.2.5.5. Vibrating Plates

When using vibrating plates, a separate vibrating unit vibrates a number of tiny plates. Vibrating plate models that can be operated by hand are useful for compacting thin layers of soil. These units are mostly used for compacting granular base courses, which have a low thickness and up to 12% fines, for highways and runways (De Groot et al., 1996; Han and Collin, 2005).

4.2.6. Drums are Conquering the Earth

Natural rock, the perfect foundation for building roads, is not always available from nature. Most of the time, the ground is unstable and doesn't provide enough bearing capacity. Therefore, the first stage of building a new road is typically earthmoving operations. The new road's foundation must be ready after the course has been determined and laid out (Figure 4.2).

- **Strengthen Resistance:** The single most crucial step in the construction of soil is compaction. Its function is to decrease the amount of air and water-filled pores in the soil that needs to be compacted. Compaction will give soil the desired qualities: it will increase soil stability and decrease its tendency to swell due to water absorption, improving soil resistance to stresses induced by climate change and traffic. Additionally, the latter will increase the soil's resistance to frost.
- **Cohesive and Non-Cohesive Soils:** For various soil types, there are differences in the process. Soils that are cohesive or non-cohesive are distinguished by experts. The soil particles form bonds with one another in cohesive soils such as loam, clay, or silt. Particles in non-cohesive soils, such as sand or gravel, lie next to one another without adhering. Water-retaining soils, fine-grained cohesive soils, coarse-grained non-cohesive soils, fine-and coarse-grained mixed soils, and rock are the different soil types based on their capacity to be compacted.
- **Dynamic Compaction is Highly Effective:** Because the soil is compacted by the combination of vibrations from the roller drum and the roller's weight, dynamic compaction with vibrating or oscillating rollers has proven especially useful in earthworks.

Figure 4.2. *Compacting a gravel layer in oscillation mode: This gentle method of dynamic compaction needs just a few roller passes to achieve high degrees of density.*

Source: https://www.wirtgen-group.com/media/01_wirtgen_group/01_bilder_1/02_anwendungen/wg_photo_jobsite-istanbul-airport_00104_hi_576x324.jpg.

- **Compaction Requires More Than Just Weight:** Three factors influence the intensity of compaction: displacement of the roller drum, which represents the amplitude of the vibration or oscillation; rate of recurrence, which represents the frequency; and duration, which represents the roller's travel speed. Correct setting of these parameters allows for the achievement of the desired density with a significantly reduced number of roller passes (Figure 4.3).

Figure 4.3. *Rubber-tired roller GRW 280: With this machine, Hamm newly defined the rubber-tired rollers featuring perfect visibility, intuitive operation and an extremely flexible ballast design.*

Source: https://www.wirtgen-group.com/media/01_wirtgen_group/01_bilder_1/02_anwendungen/2019_broschuerenmotiv_serie_grw180_280_72dpi_26_992x558.jpg.

- **The Right Technique for Any Type of Soil:** Compactors with padfoot roller drums and high amplitudes work best for compacting cohesive soils, while smooth roller drums and lower amplitudes work better for compacting non-cohesive soils. As long as there isn't still an excessive amount of moisture present, the ground is ready to be paved. If so, it needs to be stabilized in addition.

4.2.7. How Compaction is Essential to Pave Smooth Roads

The demands on city, state, and federal projects are higher than ever. Contractors are resurfacing interstates, extending highways, and constructing new roads. There will be a lot of road construction and paving as funds from the Infrastructure Investment and Jobs Act make their way to state and local governments. However, how can an unpaved road be transformed from a bumpy ride to a smooth, enjoyable drive? First, we need to comprehend how compaction impacts the smooth surfaces of roads. The process of mechanically compacting aggregate, asphalt, or soil to eliminate air pockets and boost density is known as compaction. It is an essential stage in building a road. Improving pavement strength, stability, and longevity is of utmost importance. When roads are properly compacted, they minimize the risk of premature pavement failure, simplify maintenance, and provide a smooth driving experience.

Explore the five forces that affect the compaction process to know how compaction is essential to building smooth roads:

1. Static force is the compaction equipment's weight applied to the compacted material. It is the primary force that initiates the compaction process. Vibratory, smooth-drum and tired pneumatic rollers generate static force, effectively reducing air voids and increasing density.
2. Kneading force involves manipulating the material by exerting pressure through a kneading action. This force works by rolling or kneading the asphalt mixture to achieve a uniform aggregate and asphalt binder distribution. Kneading force is particularly crucial in asphalt pavement construction, ensuring proper cohesion and reducing the potential for segregation.
3. Impact force is generated when striking the road surface with a high-energy blow. This force is commonly employed to compact granular materials and create stable foundations. Vibratory plate compactors and pneumatic tampers are examples of equipment that use impact force, effectively reducing air voids and achieving better compaction.

4. Shearing force involves moving one material layer over another, causing particles to slide and settle. It is primarily used in cohesive soils to align particles and create a denser structure. Sheepsfoot rollers and grid rollers are commonly employed to apply shearing force, enabling the compaction of cohesive soils.
5. The vibratory force uses rapid vibrations to enhance the compaction process. Vibratory rollers and compactors apply vibratory force to improve the compaction of granular materials and achieve greater density. The vibrations reduce friction between particles, facilitating the movement and rearrangement of materials, resulting in increased compaction efficiency.

When paving, operators collaborate with engineers and project managers to monitor the compaction force at each stage of the roadbuilding process using equipment technology on their pavers. Compaction is influenced by several variables, including the machine's weight, project specifications like the concrete or asphalt mix, and environmental elements like temperature and precipitation.

Contractors must monitor the five forces of compaction because their combined effect contributes to the construction of smooth roads in the following ways:

- **Improved Stability:** Compaction eliminates air voids and increases the density of materials, enhancing their strength and load-bearing capacity. When a road is an appropriate density for its environment, it prevents the formation of potholes, cracks and uneven surfaces, ensuring a smooth and safe driving experience.
- **Enhanced Durability:** Proper compaction reduces the permeability of road surfaces, preventing water infiltration and subsequent damage due to freeze-thaw cycles. It also minimizes the potential for rutting and deformation, improving the overall durability of the road.
- **Uniformity of Surface:** Compaction ensures the even distribution of materials, resulting in a uniform surface free from variations or irregularities. This uniformity improves ride quality, reduces noise and improves vehicle traction, enhancing overall road safety.
- **Increased Longevity:** A well-compacted road exhibits greater resistance to wear and tear, prolonging its lifespan and reducing the need for frequent repairs. Proper compaction contributes to cost savings and ensures sustainable infrastructure development by minimizing the potential for premature pavement failure.

A crucial component of building a road is compacting the soil, which has an immediate effect on the durability and strength of the road. Through an understanding of the theoretical underpinnings of soil behavior and the application of suitable practical techniques, engineers can guarantee the construction of robust and long-lasting roads. Compaction is a complicated but crucial process in the field of civil engineering that depends on variables like soil type, moisture content, and the use of the right equipment (De Groot et al., 1996; Han and Collin, 2005).

4.3. GEOSYNTHETICS IN SOIL STABILIZATION

Man-made polymeric (plastic) products called geosynthetics are used in construction projects to address geotechnical issues. They could be three-dimensional structures, planar, or strips. The various geosynthetic types are specifically engineered to provide a range of functions, including drainage, erosion control, reinforcement, stabilization, filtration, and liquid containment. Certain products might be made to perform multiple tasks.

It is simple to deduce the meaning of geosynthetics from the word itself. "Synthetic" refers to something created by humans, whereas "geo" refers to the ground or any land.

4.3.1. Types of Geosynthetics

There exist nine distinct varieties of geosynthetics, which vary in their conception, production, and intended applications. The categories of geosynthetics are:

1. Geogrids
2. Geotextiles
3. Geocells
4. Geonets
5. Geopipes
6. Geofoam
7. Geocomposites
8. Geomembranes
9. Geosynthetic clay liners

4.3.1.1. Geogrids

Geogrids are made up of a regular open network of tensile elements (ribs) that are integrally connected. These elements can be joined together by bonding, interlacing, or extrusion. The constituents are smaller than the apertures that separate the ribs. Polypropylene, high-density polyethylene (HDPE), and other robust polymers are examples of the polymeric materials used to make the ribs. Stretching the polymer material during the manufacturing process can help to orient the molecular structure and give the ribs more strength and stiffness.

A geogrid's robust connections and rigid ribs allow for a high level of interaction between the geogrid and the soil around it. Particles of soil can partially enter the apertures, where they are either contained by the ribs or restrained. As one of the most widely used geosynthetics, geogrids are perfect for stabilizing or reinforcing soils in applications like building over unstable soils, laying road foundations, and creating earth retaining structures (Figure 4.4).

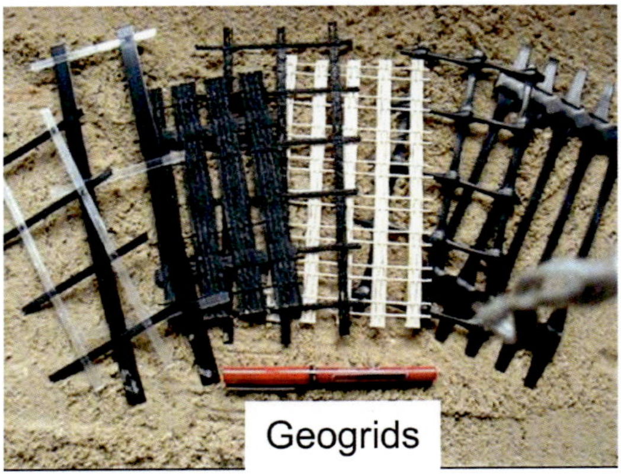

Figure 4.4. *Geogrids.*

Source: By Marilyn475 – Own work, Public Domain, https://commons.wikimedia.org/w/index.php?curid=4499781.jpg.

4.3.1.2. Geotextiles

The largest class of geosynthetics and one of the first to be developed are geotextiles. They are permeable textiles that can be knitted, woven, or non-woven. They are made of synthetic fibers like polyester or polypropylene. The directionally or randomly oriented fibers or filaments are mechanically, thermally, or chemically bonded together to create the non-woven varieties. They range in weight and strength from high-strength reinforcement materials to lightweight filter products.

Although most frequently used as a separator before road construction or as a filter/separator in drainage applications, this category of geosynthetic can provide a wide variety of functions when used in conjunction with soil, including protection, reinforcement, drainage, and separation. They can be used across a range of applications in engineering projects (Figure 4.5).

Figure 4.5. *Geotextiles.*

Source: By Marilyn475 – Own work, Public Domain, https://commons.wikimedia.org/w/index.php?curid=4482528.

4.3.1.3. Geocells

Geocells are three-dimensional geosynthetic products made of geotextile strips or webs that are ultrasonically welded together. These materials expand to form a structure resembling a honeycomb that can be filled with gravel, sand, or soil. Soil movement is reduced because the cellular structure confines and stabilizes the infill material. Because of this, geocells can be used for a variety of purposes, such as stabilization and protection against soil erosion.

4.3.1.4. Geonets

Geonets are made of a diamond-shaped net structure made of extruded polymer strands arranged in a parallel pattern. In order to establish geonets as a "drainage composite," they are occasionally laminated with geotextiles.

4.3.1.5. Geopipes

Geopipes are any polymeric pipes—perforated or solid wall—that help with liquid and gas drainage during building projects. To preserve their ability to drain, they are most frequently wrapped in a geotextile filter.

Geopipes can be used in many different contexts, such as landfill leachate or gas collection applications.

4.3.1.6. Geofoam

Geofoam, also referred to as Expanded Polystyrene (EPS), is a very strong and lightweight material that can be used in a variety of applications in place of soil backfill. By polymerizing polystyrene and expanding it, a large number of closed, gas-filled cells are produced, resulting in the formation of geofoam blocks. They have such a low density because of this design.

Geofoam's low density makes it an excellent fill material for soft or compressible foundation soils in engineering projects. When utilized as a lightweight core for an embankment, it can potentially prevent staged construction and minimize settling.

4.3.1.7. Geomembranes

An impermeable, or low-permeability, synthetic membrane is called a geomembrane. In geotechnical engineering projects, they are usually employed as a barrier for liquids or gas, regulating the flow of fluids and offering containment. Geomembranes, a geosynthetic material with chemical resistance, can be particularly helpful in situations where there is a risk of hazardous contaminants leaking. If not, they are typically employed as liners in the mining, agricultural, and landfill sectors.

4.3.1.8. Geosynthetic Clay Liners

A layer of sodium bentonite clay is layered between two non-woven geotextile sheets to create geosynthetic clay liners (GCLs). In order to create structural integrity, the sheets are bonded together (either by stitching or needle punching); the layers are then secured in place by heat treatment.

GCLs offer a quicker and more practical substitute for containment ponds' conventional clay lining. The swelling properties of the sodium bentonite layer give these materials an additional advantage. Clay liners therefore provide a level of self-sealing that lowers leakage. GCL liners are advantageous for a variety of geotechnical uses, such as landfills and waste treatment.

4.3.1.9. Geocomposites

Geocomposites are blends of two or more of the previously mentioned geosynthetic types. When the characteristics of different geosynthetics are combined, a product is produced that has more advantages than any one product type. This product is especially helpful in containment and drainage applications, as well as in some road foundation scenarios (Figure 4.6).

Figure 4.6. *Geocomposites.*

Source: By Marilyn475 – Own work, Public Domain, https://commons.wikimedia.org/w/index.php?curid=4499837.

While various polymers can be used to create geosynthetics, polyester (PET), polypropylene (PP), and high-density polyethylene (HDPE) are the most widely used ones. The chosen polymers have a high level of resistance to chemical and biological deterioration. Additives and modifiers are added to the polymer blend by manufacturers to impart desired properties like UV and weather resistance. Other polymers, like polyvinyl alcohol (PVA) or aramids, are utilized in extremely specialized applications where extreme strength is required.

4.3.2. Functions of Geosynthetics

The prefix "geo" refers to the eight primary purposes of geosynthetics in relation to earth materials, such as rocks and soils; these purposes range from liquid containment to soil stabilization. Certain products are made to perform multiple tasks, which lowers the number of products required and helps save time and money.

4.3.2.1. Stabilization

When stabilizing granular soils with geosynthetics, an interlocking mechanism is usually employed. For instance, aggregate can enter the spaces between the ribs of a geogrid and interlock, containing the aggregate material. By minimizing soil particle movement and incorporating ribs that offer high stiffness at low strain, the geogrid can enhance the mechanical behavior of the soil. A stronger and more deformation-resistant composite layer is produced by this mechanical ground stabilization.

Because geogrid stabilization increases bearing capacity and decreases deformation under load, it is frequently used in the foundations of roadways and on working platforms that will support heavy loads.

4.3.2.2. Reinforcement

By adding a reinforcing element that increases strength and resistance to deformation, high stiffness and high strength geosynthetics, like some geotextiles and geogrid types, can improve the mechanical properties of a soil mass.

Engineers can construct embankments on soft foundations and produce earth slopes with steeper slope angles by using geosynthetics that serve as reinforcement.

4.3.2.3. Drainage

Groundwater and other fluids can gather and move through less permeable soils thanks to the drainage function of geosynthetics. Geosynthetics for drainage can disperse pore pressure beneath embankments, capture groundwater on hillsides or behind buildings, and give road pavement edge drainage.

The majority of drainage geosynthetics are geocomposites, which are made up of one or more layers of geotextile and a geonet drainage core. Water, along with other liquids and gases, can be transported through their structure and collected in an open area.

Roadways require proper drainage because water beneath the surface can cause subgrade soils to soften and eventually cause the road structure to lose strength. Thus, geosynthetics are frequently seen beneath retaining walls, beneath embankments, and in areas of less permeable soil such as roads, railroads, and other structures.

4.3.2.4. Erosion Control

Reducing land damage from wind and water erosion is known as erosion control. Regrowth occurs slowly after topsoil erosion and erosion control geosynthetics help the natural process by providing a buffer against further erosion.

Erosion control geosynthetics usually come in the form of multilayered mats that lessen surface runoff and water droplet impact on soil erosion. After being rolled onto a surface, they are secured with pegs. Certain products blend natural and synthetic materials together to improve moisture retention and promote the growth of vegetation.

Erosion control geosynthetics are great for preserving the top soil layer, promoting plant growth, and averting future soil loss in areas where land is subjected to rainfall or water flow. This is especially prevalent near bodies of water and the slopes of embankments.

4.3.2.5. Filtration

Water moving through soils can carry soil particles, especially the finer ones. The purpose of filtration geosynthetics, which are primarily geotextiles, is to hold soil particles on the upstream side of the filter while letting water flow through freely. Because of the "bridging" effect of larger particles on the upstream side of the filter, even finer soil particles can be retained. Therefore, one-directional water flow is ideal for filtration.

Geotextiles' filtration qualities can be tailored by adjusting the fabric's thickness, structure, fiber type, and density. They are frequently joined in the form of a geocomposite with a drainage core. Properly designed products can be utilized for vital purposes beneath riprap protection in river or coastal works, or to stop soil migration into gravel-filled drains or drainage aggregate layers.

4.3.2.6. Separation

The geosynthetic must stop soil with varying particle size distributions from mixing together and jeopardizing structural integrity in order to serve as a separator. Although it is a necessary function in many applications, separation is crucial to the pavement and road layers. Underneath road and rail construction, geotextile separators are frequently employed either alone or in conjunction with a geogrid to create a geocomposite.

Expensive subbase material can be kept from cutting into the soft subgrade by using a geogrid. The geogrid/soil composite layer can stop finer-grained soil from migrating up into the subbase when it stabilizes a well-graded subbase. A geocomposite with separator/filter and geogrid qualities could be used in situations where soil moisture content is high.

4.3.2.7. Barrier

A comparatively impermeable barrier for the containment of fluids or gases can be created with geosynthetics. Barrier geosynthetic comes in a number of varieties. High impermeability is provided by geomembranes made of polyethylene, PVC, or rubber. An option with some self-healing potential is GCLs.

Fluid containment is a common use for geosynthetic barriers in mining, industry, and agriculture. They are also utilized in landfill capping and waste containment in conjunction with drainage composites.

4.3.2.8. Protection

To provide protection, geosynthetics can act as a cushion for a sublayer. When applying covering fill, non-woven geotextiles can be made to prevent punctures to geomembranes. Coated pipelines can be protected during backfill operations with materials of the Geonet type.

4.3.3. Geosynthetic Applications

There are many different geotechnical applications for geosynthetics. Applications of geosynthetics are the circumstances in which they can be used to provide solutions to particular geotechnical problems or challenges in an engineering project. These should not be confused with the functions covered above.

4.3.3.1. Roads, Pavements and Trafficked Areas

Applications for geosynthetics include roads, railroads, airports, and more. They are mainly helpful in stabilizing and separating unbound pavement layers for roads and runways. They can also be utilized to provide side drainage or to address problems with the underlying soil. Since the 1970s, geogrids have been used to facilitate road construction and improve performance on soft ground. Because of more recent developments, they are now more frequently used to extend the lifespan of paved roads and lower whole-life costs (De Groot et al., 1996; Han and Collin, 2005) (Figure 4.7).

Figure 4.7. *Roads, pavements and trafficked areas.*

Source: By Werner W Müller and Fokke Saathoff – http://iopscience.iop.org/1468-6996/16/3/034605/article, CC BY 3.0, https://commons.wikimedia.org/w/index.php?curid=40079980.

4.3.3.2. Rail Trackbed Improvement

Below rail track, geosynthetics can be used to address a number of issues. In areas with weak soils, stabilization geogrids are frequently used to increase the trackbed's stiffness and bearing capacity. In order to regulate the lateral migration and degradation of the ballast particles, they can also be positioned beneath the layer of ballast. With geogrid stabilized transition zones, differential stiffness problems related to the changeover from rigid to flexible foundations can be resolved.

Sand-filled geotextile mats are a highly specialized alternative to sand filters beneath trackbeds, and geotextile separators and drainage geocomposites are used to control moisture-related issues (Figure 4.8).

Figure 4.8. *Improvement fence.*

Source: By U.S. Environmental Protection Agency (EPA) – U.S. Environmental Protection Agency], Public Domain, https://commons.wikimedia.org/w/index.php?curid=4369586.

4.3.3.3. Earth Retaining Structures and Steep Slopes

A smaller footprint than natural slopes can be achieved by creating steeper slopes with the use of high strength geotextiles and reinforcing geogrids. Bridge abutments and reinforced soil walls can also be built with geogrids. Comparatively speaking, these can be built more quickly and affordably than piled or concrete retaining walls (Figure 4.9).

Figure 4.9. *Earth retaining structures and steep slopes.*

Source: By Werner W Müller and Fokke Saathoff – http://iopscience.iop.org/1468-6996/16/3/034605/article, CC BY 3.0, https://commons.wikimedia.org/w/index.php?curid=40078443.

4.3.3.4. Embankment Foundations

The application of geosynthetics can help overcome the difficulties associated with building earth embankments over unstable soils. As construction progresses, overstressing the foundation soil may cause a deep rotational failure. The stabilization of the embankment against this failure mechanism can be preserved by adding geosynthetic reinforcement to its base.

Figure 4.10. *Embankment foundations.*

Source: By Werner W Müller and Fokke Saathoff – http://iopscience.iop.org/1468-6996/16/3/034605/article, CC BY 3.0, https://commons.wikimedia.org/w/index.php?curid=40078297.

In addition to offering reinforcement at the base, three-dimensional cellular mattress systems like Tensartech Stratum® also distribute load and have an impact on the settlement profile due to their inherent stiffness. When the embankment rises, excess pore pressure can be released by installing geosynthetic wick drains deeply into the foundation soil, which speeds up construction (Figure 4.10).

4.3.3.5. Drainage Systems

Geosynthetics are well-established for water movement, filtration, and separation in drainage systems. Geocomposite materials can be used as slope interceptor drains or in place of aggregate-filled "French drains" beside roads. They also relieve pore pressure and facilitate drainage behind buried structures. The use of filter geotextiles keeps perforated drainage pipes and drainage aggregate layers clean, and prefabricated vertical drains speed up construction by relieving pore pressure beneath embankments.

4.3.3.6. Containment and Landfill

Layered geosynthetic systems made up of impermeable barriers, drainage composites, and geosynthetic protection can be used to line new landfill sites. Barriers and drainage composites will be used when capping landfill sites to allow for gas venting and containment. Tailings lagoons are lined by impermeable membranes used in the mining industry; geogrids and geocomposites are used to stabilize the capping of lagoons after they are used.

4.3.4. The Benefits of Geosynthetics in Geotechnical Engineering

Geosynthetics with proper design and application can be very beneficial. Their applicability in various applications, soil types, and climate zones has been demonstrated over time. The advantages are numerous, ranging from making it possible to cross "impossible" terrain to prolonging the life of roads. From lessening the amount of aggregate used in buildings to preventing erosion in coastal regions and waterways. Here are some of the benefits of using geosynthetics:

- **Improved Lifespan:** maintaining high performance for an extended lifetime.
- **Cost Savings:** reducing the quantities of imported fill, aggregates and waste, accelerating construction, and providing more efficient design options

- **Simplicity of Construction:** straightforward and fast installation using well-proven methods.
- **Strength:** in reinforcement applications, they have a high strength-to-weight ratio.
- **Durability:** highly resistant to weathering, biodegradation and chemical contaminants. As an example, allowing the use of lower-cost non-standard fill in reinforced soil structures.
- **Customizable:** can be tailored to the needs in terms of structure, size and composition.
- **Design Creativity:** new solutions, new construction methods and new ideas.
- **Environmentally Beneficial:** can enable alternative construction methods that reduce material volumes, transportation and construction methods – and their associated carbon emissions, to offer a net environmental benefit.
- **Resilience:** geosynthetics have enabled new and innovative solutions that help safeguard infrastructure against the effects of environmental change, such as higher rainfall, flooding, and changes to freeze-thaw weathering and degradation of roads.
- **Sustainable:** geosynthetics like geogrids can improve the service lives of structures, cut maintenance requirements and reduce whole-life costs (Stucki et al., 2011).

4.4. BLENDING AND MIXING OF SOILS FOR ENHANCED STABILITY

When building a road, soil stability is crucial because it directly affects the road's strength, longevity, and durability. Naturally occurring soils at a building site frequently lack the mechanical qualities needed to support a road's weight. In order to get around this problem, engineers frequently use methods for blending and mixing soil to increase the soil's stability and performance under large loads. These techniques modify and enhance the soil's engineering qualities, such as strength, compaction, and moisture retention, by mixing various soil types or stabilizing agents (Stucki et al., 2011).

4.4.1. The Purpose of Blending and Mixing Soils

The main goal of blending and mixing soils is to:

1. **Increase Load-bearing Capacity:** By modifying the soil structure, blending can improve its strength, allowing it to support heavier traffic loads without deformation.

2. **Reduce Compressibility:** Mixed soils have less susceptibility to compression, reducing the risk of settlement or rutting under the weight of road traffic.
3. **Improve Drainage and Moisture Control:** The addition of granular materials or stabilizers can enhance drainage, reducing the soil's tendency to retain water, which is critical for preventing the weakening of the subgrade.
4. **Enhance Durability:** Blending improves the resistance of soils to environmental factors like freeze-thaw cycles and water erosion, resulting in longer-lasting roadways.
5. **Mitigate Shrink-Swell Behavior:** In regions where soils exhibit high shrink-swell potential (like clay soils), blending can reduce these tendencies, making the subgrade more stable.

4.4.2. Blending and Mixing Techniques

4.4.2.1. Blending Different Soil Types

One of the most popular techniques for achieving a more suitable subgrade for road construction is blending different types of soil. To create a more balanced mixture, this process entails mixing soils with various particle sizes and characteristics. For instance, blending with sand or gravel can help a site whose soil is primarily clayey by enhancing compaction and drainage. On the other hand, to increase cohesiveness, highly granular soils can be combined with silt or clay.

- **Clay and Sand Mixtures:** Clay provides cohesiveness, while sand improves drainage and reduces shrink-swell behavior. This blend helps create a subgrade that is both strong and flexible.
- **Gravel and Fines:** Adding fine materials, such as silt or clay, to a gravel-dominated soil helps fill the voids between larger particles, improving compaction and reducing permeability.
- **Topsoil and Subsoil Mixtures:** In some cases, topsoil is mixed with subsoil layers to increase the fertility and stability of the soil surface, although this is more common in landscaping and agriculture.

4.4.2.2. Mechanical Mixing

Using specialized equipment, mechanical mixing is used to blend soils or mix in stabilizing agents (such as lime, cement, or fly ash). The overall qualities of the soil are improved by this process, which guarantees a uniform distribution of the materials. The kind of machinery used varies based on the size and type of soil of the project.

- **Rotary Mixers:** Rotary mixers are commonly used to blend soils or mix stabilizing agents into the soil. These machines have rotating blades that churn the soil and ensure that different soil components or additives are uniformly distributed throughout the subgrade.
- **Disc Harrows and Bulldozers:** On smaller projects, equipment like disc harrows or bulldozers may be used to blend the soil layers. These machines loosen and mix the soil, although they may not achieve the same degree of uniformity as rotary mixers.

4.4.2.3. Chemical Stabilization Through Soil Mixing

To improve soil stability, engineers frequently employ chemical stabilization techniques in addition to blending various soil types. A variety of stabilizers are added to soil to change its characteristics, including fly ash, cement, lime, and bitumen.

- **Lime Stabilization:** Lime is often added to clayey soils to reduce plasticity and improve strength. When mixed with clay, lime reacts chemically with the soil, forming calcium silicate and calcium aluminate hydrates (CAHs), which significantly enhance the soil's compressive strength and reduce moisture sensitivity. Lime-stabilized soils are less prone to shrinkage and swelling.
- **Cement Stabilization:** Cement is used in soil stabilization to bind soil particles together, creating a stronger and more rigid base. This method is particularly useful in areas where soft soils are prevalent. When cement is mixed with soil, it forms a solid, concrete-like matrix that can withstand greater loads. Cement stabilization is commonly used for both subgrade and subbase layers in road construction.
- **Bitumen Stabilization:** Bitumen can be added to sandy or granular soils to improve cohesion and resistance to water infiltration. The bitumen forms a waterproof barrier, protecting the soil from the weakening effects of water. This method is commonly used on road surfaces and in areas where drainage is poor.
- **Fly Ash Stabilization:** Fly ash, a byproduct of coal combustion, is used to improve soil strength and reduce swelling potential. When mixed with water and soil, fly ash reacts to form cementitious compounds that enhance the soil's load-bearing capacity.

4.4.2.4. Deep Mixing Methods

Deep mixing techniques are used for bigger projects or locations where weak soil goes down a considerable distance. This entails inserting mixing implements far below the surface of the earth to combine stabilizers and soil.

- **Soil-Cement Columns:** A deep mixing technique that involves mixing cement slurry into weak soils using augers to create soil-cement columns. These columns increase the strength and load-bearing capacity of the soil, improving the stability of the subgrade in areas with weak or loose soils.
- **Jet Grouting:** In this technique, high-pressure jets are used to inject stabilizing materials into the soil, creating a mixed-in-place soil-cement matrix that strengthens weak soils.

4.4.3. Applications in Road Construction

1. **Subgrade Stabilization**: Blended or chemically stabilized soils are used to strengthen the subgrade, which forms the foundation for the road. A stable subgrade minimizes deformation under traffic loads, reducing maintenance needs and extending the lifespan of the road.
2. **Base Course Construction**: The base course, placed above the subgrade, requires high strength and low compressibility. Soil blending techniques are often applied to ensure the base layer has the necessary load-bearing capacity.
3. **Pavement Foundations**: In areas with highly compressible soils, soil mixing and stabilization ensure that the pavement foundation can support heavy loads without excessive settlement or cracking.
4. **Embankment Construction**: When constructing embankments for highways or bridges, blending and mixing are used to improve the stability of fill materials. This reduces the risk of landslides or settlement in the embankment.

The stabilization and longevity of the road structure are ensured by the essential techniques of blending and mixing soils during the construction process. Engineers can overcome the challenges presented by weak or variable soils by combining different types of soil and adding stabilizing agents. Whether by chemical stabilization, mechanical mixing, or deep mixing, these methods provide efficient ways to improve the load-bearing capacity, resilience, and durability of roads in a variety of environments (Stucki et al., 2011).

REFERENCES

1. Adams, C. A., Apraku, E., & Opoku-Boahen, R. (2015). Effect of triaxial geogrid reinforcement on CBR strength of natural gravel soil for road pavements. *Journal of Civil Engineering Research, 5*(2), 45–51.
2. Al-Qadi, I. L., & Appea, A. K. (2003). Eight-year field performance of secondary road incorporating geosynthetics at subgrade-base interface. *Transportation Research Record, 1849*(1), 212–220.
3. Amu, O. O., & Adetuberu, A. A. (n.d.). Characteristics of bamboo leaf ash stabilization on lateritic soil in highway construction.
4. Anderson, R. P., Molina, F. G., Salazar, R. Z., Diaz, A. L., Sánchez, N., & Jorge, A. (2003). Geosynthetics facilitate road construction and mitigate environmental impact in Amazon Basin rainforest: 10 years of performance.
5. Anniello, P. J., Zhao, A., & Capra, G. (2003). *U.S. Patent No. 6,505,996.* Washington, DC: U.S. Patent and Trademark Office.
6. Antony, J. P. V., & Sneha, R. M. (n.d.). Effect of random inclusion of bamboo fibers on strength behaviour of fly ash-treated black cotton soil.
7. Bairagi, H. (n.d.). *International Journal of Engineering Sciences and Research Technology.*
8. Bloise, N., & Ucciardo, S. (2000). On-site test of reinforced freeway with high-strength geosynthetics. In *EUROGEO 2000: Proceedings of the 2nd European Geosynthetics Conference. Volume 1: Mercer Lecture, Keynote Lectures, Geotechnical Applications.*
9. Brandon, T. L., Al-Qadi, I. L., Lacina, B. A., & Bhutta, S. A. (1996). Construction and instrumentation of geosynthetically stabilized secondary road test sections. *Transportation Research Record, 1534*(1), 50–57.
10. Bureau of Indian Standards. (1965). *I.S: 2720 (Part XVI): Indian Standard for Laboratory Determination of CBR.* New Delhi: Bureau of Indian Standards.
11. Bureau of Indian Standards. (1980). *I.S: 2720 (Part VII): Indian Standard for Determination of Water Content-Dry Density Relationship using Light Compaction.* New Delhi: Bureau of Indian Standards.
12. Bureau of Indian Standards. (1983). *I.S: 2720 (Part I): Indian Standard for Preparation of Dry Soil Samples for Various Tests.* New Delhi: Bureau of Indian Standards.
13. Bureau of Indian Standards. (1985). *I.S: 2720 (Part IV): Indian Standard*

for Grain Size Analysis. New Delhi: Bureau of Indian Standards.

14. Bureau of Indian Standards. (1985). *I.S: 2720 (Part V): Indian Standard for Determination of Liquid Limit and Plastic Limit.* New Delhi: Bureau of Indian Standards.
15. Bureau of Indian Standards. (1991). *I.S: 2720 (Part X): Indian Standard for Determination of Unconfined Compressive Strength.* New Delhi: Bureau of Indian Standards.
16. Bureau of Indian Standards. (1992). *I.S: 2720 (Part XX): Indian Standard for Determination of Linear Shrinkage.* New Delhi: Bureau of Indian Standards.
17. Collin, J. G., Watson, C. H., & Han, J. (2005). Column-supported embankment solves time constraint for new road construction. In *Contemporary Issues in Foundation Engineering* (pp. 1–10).
18. De Groot, M. B., Den Hoedt, G., & TerMaat, R. J. (1996). *Geosynthetics: Applications, Design and Construction*. A. A. Balkema.
19. Douglas, R. A., & Valsangkar, A. J. (1992). Unpaved geosynthetic-built resource access roads: Stiffness rather than rut depth as the key design criterion. *Geotextiles and Geomembranes, 11*(1), 45–59.
20. Frischknecht, R., Stucki, M., Büsser, S., & Itten, R. (2012). Comparative life cycle assessment of geosynthetics versus conventional construction materials. *Ground Engineering.*
21. Han, J., & Collin, J. G. (2005). Geosynthetic support systems over pile foundations. In *Geosynthetics Research and Development in Progress* (pp. 1–5).
22. Han, J., & Thakur, J. K. (2015). Sustainable roadway construction using recycled aggregates with geosynthetics. *Sustainable Cities and Society, 14*, 342–350.
23. Hayden, S. A., Humphrey, D. N., Christopher, B. R., Henry, K. S., & Fetten, C. (1999). *Effectiveness of Geosynthetics for Roadway Construction in Cold Regions: Results of a Multi-Use Test Section* (Vol. 2).
24. Kawade, S., Mapari, M., & Sharana, S. (n.d.). Stabilization of black cotton soil with lime and geogrid.
25. Kinney, T. C. (1996). Use of geosynthetics in road and airfield construction in cold regions. In *Roads and Airfields in Cold Regions: A State of the Practice Report* (pp. 271–288). ASCE.
26. Kinney, T. C., & Connor, B. (1987). Geosynthetics supporting embankments over voids. *Journal of Cold Regions Engineering, 1*(4), 158–170.

27. Kinney, T., & Connor, B. (1990). Geosynthetic reinforcement of paved road embankments on polygonal ground. *Journal of Cold Regions Engineering, 4*(2), 102–112.

28. Kulkarni, V. R. (n.d.). Experimental study of stabilization of B.C. soil by using slag and glass fibers.

29. Laurinavičius, A., Oginskas, R., & Žilionienė, D. (2006). Research and evaluation of Lithuanian asphalt concrete road pavements reinforced by geosynthetics. *The Baltic Journal of Road and Bridge Engineering, 1*(1), 21–28.

30. Mittal, A., & Shukla, S. (2018). Effect of non-woven geotextile and biaxial geogrid reinforcement on the strength behavior of subgrade soil. *SSRG International Journal of Civil Engineering, 5*(10), 28–32. https://doi.org/10.14445/23488352/IJCE-V5I10P105.

31. Powell, W., Keller, G. R., & Brunette, B. (1999). Applications for geosynthetics on forest service low-volume roads. *Transportation Research Record, 1652*(1), 113–120.

32. Raghudeep, V. (n.d.). *Improvement in CBR Value of Black Cotton Soil by Stabilizing it with Vitrified Polish Waste.*

33. Santoni, R. L., Tingle, J. S., & Webster, S. L. (2001). Engineering properties of sand-fiber mixtures for road construction. *Journal of Geotechnical and Geoenvironmental Engineering, 127*(3), 258–268.

34. Stucki, M., Büsser, S., Itten, R., Frischknecht, R., & Wallbaum, H. (2011). *Comparative Life Cycle Assessment of Geosynthetics Versus Conventional Construction Materials. ESU-services Ltd.,* ETH Zürich, Switzerland, commissioned by the European Association for Geosynthetic Manufacturers (EAGM).

35. Tingle, J. S., & Jersey, S. R. (2005). Cyclic plate load testing of geosynthetic-reinforced unbound aggregate roads. *Transportation Research Record, 1936*(1), 60–69.

36. Tingle, J. S., & Jersey, S. R. (2007). Empirical design methods for geosynthetic-reinforced low-volume roads. *Transportation Research Record, 1989*(1), 91–101.

37. Tutumluer, E., & Kwon, J. (2006). Evaluation of geosynthetics use for pavement subgrade restraint and working platform construction. In *Geotechnical Applications for Transportation Infrastructure: Featuring the Marquette Interchange Project in Milwaukee, Wisconsin* (pp. 96–107).

38. Vaitkus, A., Čygas, D., & Laurinavičius, A. (2010). Use of geosynthetics for the strengthening of road pavement structure in Lithuania. In

Geosynthetics for a Challenging World: 9th International Conference on Geosynthetics (Vol. 3, pp. 1575–1580).

39. Vaitkus, A., Laurinavičius, A., & Čygas, D. (2006). Site damage tests of geotextiles used for layer separation in road construction. *Baltic Journal of Road and Bridge Engineering, 1*(1), 29–37.

40. Vaitkus, A., Šiukščius, A., & Ramūnas, V. (2014). Regulations for use of geosynthetics for road embankments and subgrades. *The Baltic Journal of Road and Bridge Engineering, 9*(2), 88–93.

41. Van Santvoort, G. P. (1994). *Geotextiles and Geomembranes in Civil Engineering.* CRC Press.

42. Vieira, C. S., & Pereira, P. M. (2016). Interface shear properties of geosynthetics and construction and demolition waste from large-scale direct shear tests. *Geosynthetics International, 23*(1), 62–70.

43. Vinod, P., & Minu, M. (2010). Use of coir geotextiles in unpaved road construction. *Geosynthetics International, 17*(4), 220–227.

44. Webster, S. L., & Santoni, R. L. (1997). Contingency airfield and road construction using geosynthetic fiber stabilization of sands (Vol. 97, No. 4). U.S. Army Engineer Waterways Experiment Station.

45. Wilmers, W. (2002). The revised German regulations for the use of geosynthetics in road construction. In *Geosynthetics: State of the Art-Recent Developments. Proceedings of The Seventh International Conference on Geosynthetics* (7(4), 22–27).

46. Yan, L., Yang, J., Gao, Y., & Liu, B. (2006). Numerical analysis of geosynthetics treatment in old road widening. *Yanshilixue Yu Gongcheng Xuebao/Chinese Journal of Rock Mechanics and Engineering, 25*(8), 1670–1675.

CHAPTER 5

USE OF RECYCLED MATERIALS IN SOIL STABILIZATION

Contents

INTRODUCTION

The stability and longevity of built structures are largely dependent on the soil, which serves as the foundational layer of construction projects. Unfortunately, weak soil strength is a problem in many areas, which is characterized by a low load-bearing capacity, a vulnerability to erosion, and compromised durability. Understanding the physical properties of soil, such as its texture, structure, density, porosity, and color, is essential to determining its suitability for construction. Structure describes how soil particles are arranged into aggregates, whereas texture refers to the proportions of sand, silt, and clay particles. The bulk of the soil per unit volume is known as density, and the void spaces that affect the flow of water and air are known as porosity. Furthermore, the color of the soil can indicate its mineral content and amount of organic matter, all of which influence how well it works as a building material overall.

Enhancing the chemical, physical, or mechanical characteristics of soil to meet the necessary requirements for strength, durability, and load-bearing capacity is known as soil stabilization, and it is a multifaceted engineering practice. In addition to being essential for accomplishing engineering objectives, this procedure is particularly important when building foundations, roads, and other construction projects.

Construction materials have changed and more unusual ingredients have been added as a result of the road industry's evolution and the increase in traffic on roads. The limited availability of traditional natural resources and the threatened environment served as the justifications for the trend toward the assessment of alternative materials and resources for integration into the road sector. Many secondary and tertiary materials are required for the inclusion of such materials. Numerous waste by-products and materials have been examined, appraised, tested for potential uses, and applied in real-world settings. The degree of inclusion varies based on the characteristics of the recycled material. Certain recycled materials have been shown to have better qualities than others and to work well in practical applications. Nevertheless, several issues with this integration have been raised by field and laboratory experiments, which have shown to be crucial for more in-depth research. Among the lengthy list are recycled concrete aggregates, plastic wastes, scrap tires, coal combustion products like fly ash, bottom ash, and pond ash, steel slag, oil sand, oil shale sand, lateritic soil, and reclaimed asphalt pavement (RAP). Using secondary and tertiary materials in road construction is thought to result in a magnificent preservation of valuable natural resources. Nevertheless, some performance-related problems would not be fixed and would persist if there was not strict collaboration between the academic community and the business sector, as well as the education of those who frequently engage in paving activities.

5.1. THE USE OF RECYCLED MATERIALS

The use of fully recycled or partially recycled materials in the production of a vast range of products is becoming more common as a means of improving sustainability and lowering environmental impact. Any plastic product, in particular, is likely to have a version on the market with at least some recycled material. Using recycled materials reduces the need for virgin materials and eliminates waste, which are its two main advantages. But there are additional considerations that must be made; these will be covered in this section (Figure 5.1).

Figure 5.1. *Recycled materials.*

Source: By © 2008 by Tomasz Sienicki [user: tsca, mail: tomasz.sienicki at gmail. com] – Photograph by Tomasz Sienicki (Own work), CC BY 3.0, https://commons. wikimedia.org/w/index.php?curid=7241890.

5.1.1. What Is Recycled Material?

When a material is "recycled," it means that it has undergone a manufacturing process to transform it from reclaimed material into a final product, material, or component that can be used in another product. Recycled materials are classified into two groups by BS EN ISO 14021:2016: either they are made from i) pre-consumer waste (any material diverted from the waste stream during manufacturing, like cutting waste), or ii) post-consumer waste (household or business waste) that can no longer be used for its intended purpose, like plastic bottles that are no longer needed.

Observing what cannot be categorized as "recycled" is interesting. The term "recycled" does not apply to rework, regrind, or scrap produced in a specific process that can be recovered within the same process that produced it. This would include, for example, sprue and flash from the molding process made of thermoplastic rubber (TR), thermoplastic polyurethane (TPU), or polyvinyl chloride (PVC) that could be chopped or ground and reinserted into the molding process (Austoroads, 2009).

5.1.2. Six Unexpected Uses for Recycled Materials

The United States recycled 69.1 million tons of municipal solid waste in 2018. These six unconventional uses for recycled materials may surprise you. However, have you ever considered where these recycled materials end up?

5.1.2.1. Art

Recycled art genuinely transforms trash into treasure for the creator. The concepts of recycled art are material reuse and environmental preservation. It really doesn't matter what is used or how the pieces turn out as long as the materials that were used to create them were discarded. Pablo Picasso was the first artist to use repurposed materials to create new works of art; Picasso was a specialist in collage. He would combine various pieces of paper, newsprint, and other materials to generate an original picture.

Figure 5.2. *Recycled cans.*

Source: By Victorgrigas – Own work, CC BY-SA 3.0, https://commons.wikimedia.org/w/index.php?curid=19886326.

In recent decades, the Recycled Art Movement has grown in popularity in tandem with the increased demand for recycling. An example of an annual event is the Recycled Art Exhibition in Flagstaff, Arizona, wherein participation is limited to items that are constructed with a minimum of 80% recycled or repurposed materials. In addition to reducing waste, painting with plastics strongly conveys to spectators the advantages of conservation (Figure 5.2).

5.1.2.2. Phones and Laptops

Instead of ending up in a landfill, your old electronics can be recycled and used to make new phones, laptops, and other devices! Some old electronics that are donated to an electronic recycling facility are stripped for raw materials because they are too old or broken to be used again. On the other hand, some recycled electronics simply require some TLC from businesses like Regency Technologies to make them into entirely new products.

This is accomplished by removing and replacing any damaged parts, testing the machine multiple times to make sure it is operating flawlessly, and finally wiping the memory to protect user privacy. These products are listed for sale online after they have undergone extensive repair and refurbishment (Figure 5.3).

Figure 5.3. *Recycled electronics to be used for new phones, computers, and more.*

Source: By Ondřej Martin Mach – Ekolist, CC BY-SA 3.0, https://commons.wikimedia.org/w/index.php?curid=11154943.

5.1.2.3. Olympic Medals

Recycled electronics can be used to create new electronics, such as computers and phones, but they can also be used to make Olympic medals! The Tokyo 2020 Summer Olympics featured gold, silver, and bronze medals that were made entirely of recycled electronics from consumers. Donations of outdated laptops, digital games, cameras, and phones were requested from the Japanese public by the Tokyo Organizing Committee. With an astounding 158 million pounds of electronics donated—roughly 7,700 pounds of silver, 4,850 pounds of bronze, and 71 pounds of gold—the project's success is evident. The gold, silver, and bronze were taken out of the electronics by highly skilled contractors who then sorted, disassembled, and melted them down to create medals (Figure 5.4).

Figure 5.4. *Recycled aluminum can art.*

Source: By Elekes Andor – Own work, CC BY-SA 4.0, https://commons.wikimedia.org/w/index.php?curid=72820388.

5.1.2.4. Water Pipes

There are 2.2 million miles of subterranean pipelines in the United States alone that provide millions of people with drinking water. Fortunately, some of these pipes are now constructed from recycled materials. That's enough pipes to wrap around the Earth 88 times. For instance, with businesses such as US Pipe, ductile iron pipes can be produced with 98% recycled material. The iron and steel used in these pipes is scrap metal, which comes from a wide range of items including automobiles, washing machines, ships, and bridges. Potable water and sewage distribution are two common uses for ductile iron pipes. Apart from ductile iron pipes, over 40% of recycled HDPE resin can be used in the production of high-density polyethylene (HDPE) pipes, which are mainly utilized for agricultural drainage and subterranean stormwater infrastructure. HDPE resin can be found in a variety of everyday household items that are recycled, including milk jugs, grocery bags, shampoo bottles, and toys. Between 1,600 and 2,200 used bottles are needed to create a single standard 20-foot length of 48-inch diameter HDPE pipe.

5.1.2.5. Playgrounds

Certain companies also utilize HDPE materials—most notably, milk jugs—in playground equipment found in parks across the nation. For instance, Superior Playgrounds uses only 100% recycled post-consumer plastic to make its decks, posts, and barriers. Compared to conventional wood playgrounds, these recycled parts are far stronger because they are reinforced with fiberglass components. They also have protection against fungus, termites, oil, and salt spray, as well as against corrosion, rotting, and cracking. Another example comes from the business Karkat, where about 95% of their recycled playgrounds are constructed from recycled materials. They can make one pound of lumber, enough to build a playground, for every eight milk jugs that are recycled.

5.1.2.6. Stadium Seats

8 million metric tons of plastic are discarded into our oceans annually. Although recycling this plastic can be costly and difficult, it is essential to protect our marine environments. One way to start reducing the amount of plastic in our oceans might be to replace the seats that fans have in stadiums. Owing to the severe deterioration that the chairs face, thousands of tons of physical waste are produced when they are replaced several times during the venue's lifetime. Businesses like Sea Chairs are working to address this. The first stadium seats ever made from recycled ocean plastics were made by Sea Chairs. Everton, a

soccer team from Liverpool, intends to install this technology by 2023 in order to reduce the environmental impact of their new stadium, even though the seats are not in use at the moment (Figure 5.5).

Figure 5.5. This upcycled cutting board was made from boxcar floorboards that were taken from a railroad salvage yard.

Source: By Hu Nhu – Own work, CC BY-SA 4.0, https://commons.wikimedia.org/w/index.php?curid=107532982.

5.1.3. Examples of Typical Recycled Materials

Numerous recycled materials and components have been introduced in almost every industry in recent years. Everything from toys and office furniture to clothes and shoes has recycled plastic waste from sources like used plastic bottles and recovered ocean plastic. The majority of paper and card packaging products are also frequently made, at least in part, from recycled materials (Austoroads, 2009).

5.1.4. Legislation

Industry-specific laws governing the labeling of materials and compositions in various products will differ, but overall, there doesn't appear to be much in the way of legislative guidelines concerning the communication of recycled content use. To make sure that any sustainability-related claims made on products marketed to consumers are reliable and withstand close examination, laws are being tightened globally. The European Union is expected to launch its own Green Claims Initiative later in 2022. For example, the US "Guides for the Use of Environmental Marketing Claims" (16 CFR 260), the UK's "Green Claims Code," and equivalents in other countries, offer guidance about what can and cannot be claimed about a particular product in any supporting documentation or marketing materials.

5.1.5. Verification of Recycled Materials

Businesses are becoming more and more interested in being able to confirm that the recycled materials they use in their products are, in fact, recycled. In addition to providing credibility to any claims made, this boosts supply chain confidence and helps businesses stand out in the marketplace. It also helps prevent accusations of "greenwashing," which is the practice of making false claims about an organizations or product's environmental credentials (Austoroads, 2009).

5.1.6. Challenges to Be Faced

Although using recycled materials can lessen a product's environmental impact, there are some potential risks and difficulties that a business should be aware of.

Making sure recycled materials are appropriate for that application and can reliably meet the same performance requirements as standard materials may be the most crucial factor to take into account when using them. Regardless of any environmental claims made, the materials used should be suitable, safe, and offer sufficient longevity, performance, and durability.

The laws governing prohibited ingredients in various products are regularly updated. Therefore, it is essential to make sure that any materials containing recycled content adhere to the relevant laws and that recycling historical materials containing substances that are either prohibited or not allowed in the item into which the material is being recycled does not cause any problems.

The environmental footprint is an additional factor. This specifically refers to the energy used in gathering waste, moving it to a recycling facility (which may be located halfway around the globe), and then repurposing it into new materials. Such energy use will offset some of the environmental advantages of using recycled materials, to varying degrees. It is important to take into account a material's entire environmental impact rather than just its recycled content.

In order to ensure a steady and dependable supply of recycled materials for use as production inputs, recycling processes must also produce a steady and dependable supply of waste, which is necessary for the manufacturing of those materials. The COVID-19 pandemic presented logistical challenges that occasionally hindered the waste's supply chain. It may become more difficult to find a supply of some plastics to recycle into pellets, which can then be used to make materials for a variety of applications as well as finished products, as laws all over the world restrict or tax the production of plastic bottles and other single-use plastics.

Some of the more widely used recycled materials, like polyester, are also coming under more scrutiny. This is seen, for instance, in a recent communication from the EU regarding a strategy for sustainable and circular textiles. Since the process used to turn plastic bottles into textiles lowers the quality of the polymer, it is better to refer to this as "downcycling" as opposed to recycling. Furthermore, the infrastructure for recycling "fiber-to-fiber" is scarce, making it very challenging to recycle a polyester-based material using current technologies if the polyester has been mixed with other fiber types. Less than 1% of textile waste is thought to be recycled into new textiles worldwide.

Lastly, variations in the materials used as inputs in the recycling process (such as color) may occasionally result in variations in the final product's appearance, which might not be as consistent or uniform as traditional material. Even though these materials are internal components and are not visible in the finished product, everyone involved in the supply chain needs to understand this and make sure they are not rejected for aesthetic reasons. A variation like this might even be promoted as a way to showcase the product's "environmental credentials."

5.1.7. Recycled Materials Versus Recyclability

It is crucial to remember that adding recycled materials to a product does not automatically make it recyclable. Businesses that choose to use recycled materials need to be cautious to distinguish between claims of recycled materials and recyclability.

There are several factors to take into consideration, and even though the individual materials or components used to make an item can be recycled, it might not be possible to recycle the item itself or at least not easily. First, can an item really be disassembled into its component parts and is the technology in place to do so feasible if it is made of several different materials or components that would need to enter different waste streams to be recycled?

Simpler products composed of a single kind of plastic might be recyclable using standard home recycling programs. They might be excluded from that waste stream if they are not properly labeled, which is necessary to make sure the recycling facility can quickly identify them.

5.1.8. The Future

The need to find ways to turn waste into new materials will increase as businesses work to adopt a circular economy model. Additionally, it will become typical for products to be created with a specific "end-of-life" solution in mind. This could entail using fewer different kinds of materials or making sure that it is simpler to disassemble into its component pieces for recycling or disposing of waste.

There is already a rise in the number of businesses creating their own take-back programs to reprocess used goods, either independently or in tandem, across several industries. If feasible, the item can be disassembled to guarantee that every material is transferred to the best end-of-life solution or refurbished for resale (Austoroads, 2009).

5.2. IMPORTANCE OF WASTE RECYCLING

Using recycled materials is primarily motivated by the depletion and scarcity of natural resources as well as the necessity of large quantities of materials for the construction of pavement layers. One way to sum up the advantages of using recycled materials as alternative ecomaterials is as follows (Kandhal and Mallick, 1998):

- Reduce wastes
- Reduce construction costs
- Save natural resources
- Reduce energy required for the production of pavement construction materials
- Save energy and reduce emissions as recycled materials are less energy-intensive in processing and hauling compared to the natural materials
- Preserve the environment
- Reduce the amount of waste dumped in landfills
- Diminish carbon emissions

When recycled materials and by-products are used effectively in road construction applications, human toxicity potential, energy use, emissions, and pollutant leaching into the land are minimized, leading to sustainability and environmental benefits. Furthermore, there is less potential for global warming, less energy and water used, and less hazardous waste.

5.3. USE OF RECYCLED MATERIALS IN ROAD CONSTRUCTION

The development of the road industry and the massive increase in the number of vehicles on the road have served as justifications for encouraging the use of all practical resources to construct better roads with longer lifespans. In the 1980s, as conventional raw materials like bitumen, crushed aggregates, and unbound aggregate mixtures became too scarce, the use of unconventional construction materials in road construction began. Because natural resources are being severely depleted, the process of producing aggregate materials has been severely disrupting both the environment and the economy. Moreover, the efforts to look into the possible integration of different by-products in road construction are motivated by the startling rate of increased waste production. Plastic wastes, scrap tires, foundry sand, oil sand, bottom and fly ashes, marble dust, recycled concrete aggregates, RAP, and steel slag are currently the most widely used recycling materials. Numerous studies have been carried out in this area. Additionally, field research and ongoing studies are being conducted to find better ways to use these recycled materials. Environmental concerns and a lack of practical field experience have hindered the use of recycled materials, despite the advantages, promising research recommendations, and practical results that come from substituting waste materials for natural raw materials. Certain waste byproducts and recycled materials have more desirable qualities than others, and when they are properly described and combined with other building materials, they can provide significant advantages.

It has been reported that various recycling materials are used in different layers of the road structure, ranging from the top surfacing layer to the layers beneath it. What defines and encourages the use of that particular recycled waste material are the processing costs, the engineering properties, the proof of the material's viability, and its favorable effect on the long-term performance of the road construction works. To develop a suitable and feasible use in terms of material type and applied dosage, it is imperative to thoroughly comprehend the behavior of any recycled by-products and explore the impact on the suggested inclusion with construction material.

The performance of asphalt pavement under varying traffic loads is affected by the shortcomings and shortcomings in the quality of the asphalt binder produced by domestic crude oil producers. Another factor supporting the search for a replacement has been the cost and quality of asphalt cement used in the road industry. Concerns about the use of refined motor oil with asphalt cement in Ontario provide evidence in favor of the bitumen-related issue. According to reports, roughly half of the used asphalt cement used in road construction contains remnants of material from the engine motor oil refining process. This is a terrible course of action that will change the asphalt cement's rheological characteristics and result in poor HMA performance. The road paving industry has suffered from shortages and scarcity of natural quality aggregates, high traffic loads imposed, and harsh environmental conditions. The goal of studies on current recycling materials was to increase the value of using recycled materials in road building and enhance asphalt pavement's long-term performance.

Any effective use of recycled materials in road building projects is focused on attaining a sustainable pavement construction. The industry is still hesitant to use the currently recommended materials, according to the Ministry of National Resources' report on the state of aggregates in Ontario. This reluctance is attributed to a lack of relevant field and practical experience as well as some shortcomings in the performance of the materials that have been put into use. According to the 2009 Ministry of National Resources of Ontario's State of Aggregates Report, recycled aggregates accounted for only 18 to 19% of the aggregates required annually for road construction, or a total of 130 million tons (Austoroads, 2009).

5.3.1. Recycled Materials and By-Products in Road Construction

The main goals of using waste materials in road construction are to lessen the negative environmental effects of processing natural materials, to relieve the pressure on governments in developed and developing nations to provide landfills and make provisions for such wastes, and to reaffirm the industry's commitment to improving road services and ride quality. Furthermore, utilizations are also rationalized intuitively by the scarcity of natural resources. Numerous experimental and research studies have been conducted to explore the possibility of utilizing waste materials in the field of road construction. Numerous studies have demonstrated the effectiveness of recycling and reusing some of these waste materials' compositions in pavement structures, and comprehensive research studies are currently being conducted on others to further illuminate the benefits of recycling these materials into pavement constructions.

Waste materials are a major source of concern for governmental organizations and transportation law authorities around the world due to strict knowledge requirements and a lack of funding for in-depth analyses of using certain waste materials most advantageously. In order to reduce greenhouse gas (GHG) emissions and preserve natural resources, international cooperation is essential. This is achieved through utilizing viable recycled materials in the construction industry and putting emphasis on evaluating methods for using them (Figure 5.6). A non-conventional material is one that doesn't meet the traditional specifications for the necessary properties. The need for secondary materials and waste by-products to be included in the road construction field has become urgent due to the constant demand for aggregates in the field, the lack of approved quality materials, and the desire to protect natural resources. Numerous recyclable waste materials have been incorporated into projects worldwide, demonstrating the potential to save costs while simultaneously safeguarding the environment.

The behavior and pattern of these materials, as well as their transformation effects when used alone or in combination with other materials in the structure of road pavement, are fundamental factors that affect both the short- and long-term performance of roads. Therefore, it is imperative to fully understand these materials' characteristics. It is thought that the enormous demand for natural material substitutes in road construction has been effectively met, as evidenced by the countless studies devoted to carrying out viability assessments on employing these substitute materials while taking sustainability into account from both an environmental and financial standpoint.

Large amounts of solid waste and byproducts were produced as a result of the urbanization and significant population growth. Previously, landfills were used to dispose of these wastes, which degraded the environment. Around the world, new roads are being constructed and the current ones are being kept up. This calls for massive quantities of natural resources, which are becoming harder to find and more costly to mine and process. It also necessitates massive energy consumption during processing and transportation, which further degrades the environment. For sustainable, financial, and environmental reasons, it is, therefore, necessary to partially substitute alternative materials, such as waste or by-product materials, for the natural conventional materials used in the construction industry.

Over 4.5 billion tons of waste are produced annually in the United States, according to estimates. Merely 500 million tons of these wastes are recycled and repurposed as recycled by-products, which include used tires, slags, coal consumption products, RAP, and materials for construction and demolition (C&D). Every year, Canada produces a significant amount of waste from construction.

With an average of 5 t per person annually, the total amount of waste generated in Europe in 2014—across all households and economic activities—was over 2.5 billion tons for the 28 European Union member states. Only 2.3 billion tons of wastes were treated by recovering or disposing of them; recycling made up roughly 36% of all treated wastes. Australia produced roughly 66 million tons of waste in total in 2010–2011, including fly ash from power plants. Each capita's share was roughly 2.8 t/year. The generated waste amounted to approximately 48 million tons (2.2 t per capita annually), with an average rate of resource recovery of roughly 60%. Figure 5.7 displays the total amounts of waste generated according to the category of recycled materials, which includes fly ash. Asia produces roughly 160 million tons of waste annually, the majority of which come from China because other nations' data are lacking (Table 5.1).

Table 5.1. *Waste Treatment Classification in Japan (Bakis et al., 2006)*

Waste Type	Classification	Description
General Waste	Household waste	Items originating from the everyday activities of ordinary households.
	General business waste	Items not classified as industrial waste but produced from commercial operations.
	Human waste	
Industrial Waste	20 legally stipulated varieties of items that emerge from commercial activities.	Items associated with all business activities (1–12): cinders, sludge, waste oil, waste acid, waste alkali, waste plastics, rubber scraps, metal scrap, glass/concrete/ceramic scraps, mine slag, debris, dust.
		Items for which the types of industries that emit emissions are limited (13–19): waste paper, wood shavings, fiber waste, animal-based solid waste, animal and plant residues, animal excrement, and animal corpses.
		(20) Concrete solidified sludge, etc., that has been treated for the purpose of disposing of industrial waste listed in (1) to (19), but does not fall under (1) to (19).

Specially Controlled Waste	Specially controlled general waste	Among general waste, there are specially designated hazardous items: PCB-containing items such as air-conditioners, televisions and other. Ash collected from garbage incineration facilities and general infectious disease waste discharged from hospitals and other locations.
	Specially controlled industrial waste	Among industrial waste, specially controlled items: (1) Waste oil (flammable waste oil) (2) Waste acid (waste strong acid) (3) Waste alkali (waste strong alkali) (4) Infectious waste (5) Specified hazardous industrial waste • Waste polychlorinated biphenyls (PCBs), etc. • Polychlorinated biphenyl (PCB)-contaminated materials • Polychlorinated biphenyl (PCB)-treated products • Specified sewage sludge • Mining slag • Waste mercury, etc. • Waste asbestos, etc. (asbestos) • Waste oil (waste solvents) • others

Source: Moshkal M, Akhapov Y, Ogihara A. (2024). Sustainable waste management in Japan: challenges, achievements, and future prospects: A review. Sustainability, 16(17):7347. https://doi.org/10.3390/su16177347 Leticia Myriam Torres Martínez, Oxana Vasilievna Kharissova, and Boris Ildusovich Kharisov. Handbook of Ecomaterials. Springer. https://doi.org/10.1007/978-3-319-68255-6.

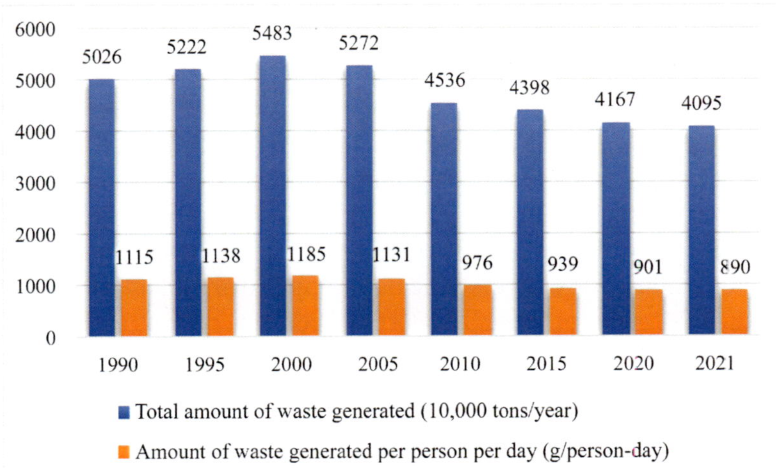

Figure 5.6. *Trends in waste generation.*

Source: Moshkal M, Akhapov Y, Ogihara A. (2024). Sustainable waste management in Japan: Challenges, achievements, and future prospects: A review. Sustainability, 16(17):7347. https://doi.org/10.3390/su16177347.

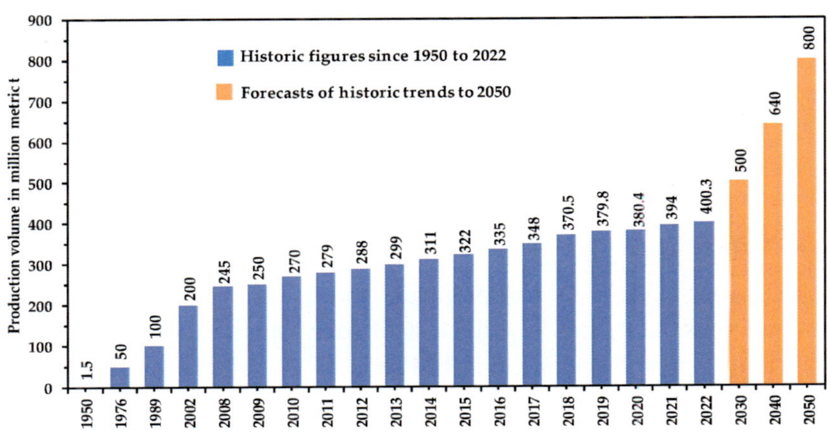

Figure 5.7. *Cumulative plastic production in million metric t.*

Source: Macheca AD, Mutuma B, Adalima JL, Midheme E, Lúcas LHM, Ochanda VK, Mhlanga SD. (2024). Perspectives on plastic waste management: challenges and possible solutions to ensure its sustainable use. Recycling, 9(5):77. https://doi.org/10.3390/recycling9050077.

5.3.2. Types of Recycled Materials in Use for Road Constructions

Recycled materials and by-products can be used to replace nonrenewable virgin aggregates and lower energy consumption in road construction, as this process consumes large amounts of natural resources and energy. Pavements made of recycled materials can be either rigid or flexible. All pavement layers, whether they are rigid or flexible, may have been built using recycled or waste materials. Waste and byproducts can therefore be used to improve asphalt mixes, whether they are hot, cold, or warm, as well as Portland cement concrete (PCC), unbound granular materials, and subgrade soils. The ensuing subsections provide an overview of the uses of recycled materials and byproducts in road building.

5.3.2.1. Flexible Pavements

Reclaimed concrete aggregate (RCA), RAP, and blast-furnace slag can be substituted for the virgin aggregates in the asphalt mix (either hot, warm, or cold). Hot mix asphalt (HMA) applications can also make use of fly ash, cement dust, sawdust, and crumb rubber. But in PCC as opposed to HMA, steel slag, fly ash, and RCA are more prevalent.

5.3.2.2. Embankments

Rice straw, compost, and leftover pulp and paper products can all help stabilize the uppermost layer of a roadside embankment by enhancing soil quality, promoting plant development, and lowering runoff. Coal ash, steel slag, and foundry sand are suitable materials for embankment fill as well.

5.3.2.3. Unbound Granular Materials (Bases/Subbases)

Unbound granular materials can be made from a variety of recycled and by-product materials, including fly ash, blast-furnace slag, steel slag, glass, roofing shingles, bottom ash, foundry sand, RAP, and RCA. Table 5.2 lists the various recycled asphalt materials.

Table 5.2. *Summary of Various Research in Terms of the Type of Method Used, Materials, Asphalt Type, Mixture Proportions, Preparations of Mixtures, and Evaluation Tests*

Waste Polymer	**Polymer Content (by Weight of Blend)**	**Asphalt**	**Mixing Method**	**Mixer**	**Mixing Parameters (Temperature, Speed, Time)**	**Tests**
HDPE and PP	0.5–3%	bitumen AC-20	Wet	High shear blending	160–170°C, 120 rpm, 30–60 min	FTIR spectroscopy Basic rheological parameters
Recycled PE	15%	asphalt modified with GTR/PE	Dry	High shear mixer	180°C, 5,000 rpm, 60 min	Temperature behavior Storage stability Morphology
PE of various densities	5%	base asphalt (70/100)	-	High shear mixer	150–170°C, 500–4,000 rpm, 30–150 min	Storage stability Morphology Fluorescence microscopy
PE (polyethylene waste packing tape)	2%, 4%, 6%, and 8%	base bitumen (60/70)	Wet	High shear blending	180°C, 20,000 rpm, 60 min	Rutting and fatigue Rheological characterization

Composite, PE, and CR	PE (6%, 7%), CR (11%, 15%)	70# base bitumen	Dry	High-speed shear process	160°C, 5,000 rpm, 60 min	Physical and rheological performance Multiple Stress Creep Recovery (MSCR) Bending-Beam Rheometer (BBR)
wPET/ GTR	3%, 5%, 7%	road bitumen (60/70)	Dry	High shear mixer	150°C, 350–3,000 rpm, 15–45 min	Rheological properties FTIR spectroscopy Morphology Thermal stability
PP, LDPE, and HDPE	2%, 4%, 6%, 8%	refinery asphalt	Wet	High shear mixer	160–190°C, 5,000 rpm, 50–60 min	Viscoelastic performance Resilient modulus Rutting, creep, and fatigue performance
EVA copolymer	3%, 5%, 7%	blown asphalt	Wet	High shear mixer	150°C, 1,500 rpm, 180 min	Physico-chemical characteristics
Various-density PE	2% + 3% SBS	base bitumen C170	Wet	Mechanical shear mixer	180°C, 3,500 rpm, 180 min	Rheological behavior Thermal behavior Conventional bitumen test

PE	0.2%, 0.3%	70# base bitumen	Dry	-	160°C, 5 min	Uniaxial penetration, wheel tracking, and four-point bending
PET	2%, 4%, 6%, 8%, 10%	bitumen (60/70)	Dry and wet	High shear laboratory-type mixer	155–160°C, 1,100 rpm, 180 min	Susceptibility Test, Indirect Tensile Fatigue Test Flow and Marshall Stability Tests
rPW, rLDPE, rHDPE	2%, 4%, 6%, 8%	sulfur-extended asphalt	Wet	high shear mixer	50–145°C, 10 min	Advanced rheological test Basic and rheological properties
PBR	5%, 7.5%, 10%	RAP materials (60/70)	Wet	high shear mixer	160°C, 6,000 rpm, 60 min	Moisture susceptibility Fatigue resistance, rutting resistance
PTP	2, 4, 6, 8, 10, 12%	asphalt binder (60/70)	Dry	high shear mixer	140–300°C, 180 min	Basic and rheological properties, Marshall, indirect tensile, and rutting tests
CSR copolymer	2%	hot-mix epoxy asphalt binder	Wet	high shear mixer	150–160°C, 200 rpm, 30 min, 180 min, 3d	Mechanical and tensile tests Thermogravimetric analysis Viscous measurement microscopy

PBR SBS	3.0%, 4.0%, 5%	bitumen (50–70)	-	high shear mixer	140–160°C, 4,000 rpm, 40 min	Preliminary Characterization Tests Chemical characterization Advanced Characterization Tests
PE	5%	asphalt (70/100)	Dry	high shear mixer	170°C, 3,500 rpm, 60 min	Semi-circular bending test cyclic compression test microscopy

Source: Jwaida Z, Dulaimi A, Mydin MAO, Özkılıç YO, Jaya RP, Ameen A. (2023). The use of waste polymers in asphalt mixtures: Bibliometric analysis and systematic review. Journal of Composites Science, 7(10):415. https://doi.org/10.3390/jcs7100415.

The use of waste and recycled materials in pavement construction has a substantial impact on the environment and economy of both developed and developing nations, as it is one of the major consumers of natural resources. However, by working together, academia and industry can improve pavement performance and address performance issues that arise from substituting secondary (recycled) materials for primary (virgin) materials. Numerous studies examined these kinds, their suitability for use in various pavement layers, the performance issues and/or gains they brought, as well as ways to address them. The following paragraphs provide an overview of the performance-related features of some of the materials mentioned above:

5.3.2.3.1. Reclaimed Asphalt Pavement (RAP)

RAP has been reported to make up as little as 8% of all recycled material used in Ontario's base material and HMA between 2004 and 2006. The characteristics of RAP differ between sources based on the kind and attributes of the aggregate material, the amount and stiffness of the binder, and the scarified pavement. One important factor affecting the performance of HMA is thought to be the concentration of RAP in the new asphalt mixture. According to reports, an asphalt pavement composed of new material would perform just as satisfactorily as one composed of a low to intermediate rate of 25% RAP. Technically, when RAP makes up less than 15% of the mixture weight, adjusting the binder grade in the asphalt mixture—particularly the low-temperature grade—does not pose a significant risk. According to reports, a new asphalt pavement's performance would be on par with one made of a low to intermediate rate of 25% RAP. When a rate of more than 20% of RAP is used, the consequences for the binder grade are much greater than when the rate is in the low range (< 20%). According to published research, if there is 15–25% RAP in the mixture, the grade must be lowered by 6°C at both low and high temperatures. However, the low-temperature grade of asphalt cement increases by 6°C when an average rate of 20% RAP plus 3% RAS is used. As a result, a decrease in the low-temperature performance grade of asphalt cement was recommended. According to the requirements of the MTO specification, the maximum allowed rate of RAP in an HMA application is 30%. Premature road failure is thought to be caused by higher added RAP in HMA applications; this needs to be investigated by routine testing on HMA that contractors supply. Simon Hesp claims that contracts frequently use higher rates of RAP than are allowed, which can destroy a million-dollar investment for a few 1,000 dollars' worth of savings. He also believes that industry education is essential to the successful implementation of RAP in HMA.

Table 5.2 shows that RAP is suitable for use in base and asphalt concrete mixtures, including warm mix asphalt (WMA), cold asphalt mixes, and HMA. The most cost-effective application of RAP, however, is in asphalt mixtures because the old asphalt in the RAP will lower the amount of binder needed in the finished mix. The costliest ingredient in the mixture is the asphalt binder. Similar performance to virgin sections was observed in a survey of Long-Term Pavement Performance (LTPP) sections across the United States and Canada that had at least 30% RAP. RAP-asphalt mixes with up to 35% RAP showed that recycled mixes outperformed conventional HMA in terms of resilient modulus and moisture susceptibility. According to a study done in the State of New Jersey to assess the use of RAP in base and subbase applications, RAP's field elastic modulus and laboratory resilient modulus are comparable in strength to dense graded aggregate base and subbase material. In a recent study, various RAP-virgin aggregate blends with RAP percentages of 0, 20, 60, 80, and 100% by total weight of aggregate mass demonstrated performance that was on par with or even superior to that of the virgin aggregate base material.

5.3.2.3.2. Recycled Asphalt Shingles (RAS)

There are two types of asphalt shingles: organic and fiberglass shingles. Both varieties contain different amounts of hard rock granules (very small, ceramic-coated particles), mineral fillers, and mineral fibers in addition to asphalt cement, which makes up an average of 30% of the total mass. Recently, the Canadian province of Ontario has started to produce organic shingles. These shingles are coated with premium aggregate granules and have a harder bitumen type than what is typically used in HMA works. However, the composition of the asphalt shingles produced by each manufacturer would differ somewhat. Because fiberglass shingles have an asphalt cement content that is approximately 15% lower than that of organic shingles, they are more affordable and desirable than organic shingles. RAS have between 19 and 31% asphalt cement by weight. When building roads, the goal of using asphalt shingles is to use less virgin binder.

Post-manufacturing, or pre-consumer, shingles are the term used for manufacturing scrap and tear-off scrap shingles, respectively. When considering recycled shingles for road construction projects, it's important to keep shingles from various sources sorted for appropriate design and implementation. The range of AC content in the RAS from various sources that were considered to be highly significant varied from 14.6 to 36.7%. Because of aging and environmental exposure, the asphalt binder from tear-off shingles is thought to be stiffer than the binder from pre-consumer shingles. It has also been reported to have improved tensile strength, resistance to rutting, and resistance to moisture damage. Nonetheless, attention needs to be paid to the resistance to fatigue and thermal cracking. Because RAS are made of an air-blown asphalt binder that is notable for its highly viscous property compared to its virgin counterpart used in the hot mix asphalt mixture, their use is restricted to 5% of the mixture's total weight. Because of the material's aging and brittleness, which have been shown to negatively affect asphalt pavement's performance at low temperatures, there are several concerns regarding the incorporation of RAP and RAS in asphalt pavements. The study recommended more research be done on this matter for the OGFC mixture because the RAS-modified mixture performed poorly at the lower temperature. The only method of field evaluation was visual.

When using RAS, it is advised to use less stiff asphalt cement to overcome the thermal cracking deficiency. The percentage of post-manufacturing or tear-off shingles that was found to be adequate to produce an asphalt mixture that met specification requirements was 5%. To create a RAP-modified asphalt mixture that is more resistant to rutting, fatigue, and thermal cracks, however, just 3% of recycled shingle needs to be added to the mixture. The addition of RAS to asphalt mixtures at different rates (5% to 10%) showed that resistance to thermal cracking and tensile strength decreased as RAS concentration increased. RAS is thought to increase the durability of HMA. When 5 to 10% of asphalt shingles were used, the same conclusion was reached. One of the very few studies on the use of RAS and RAP in asphalt pavement in Canada concluded that RAS had a greater influence on the properties of asphalt mixture than RAP, and that the mixes under investigation performed better when 3% less RAP-modified HMA was used following Ontario's requirements for asphalt pavement. On asphalt pavements that had been altered with a high rate of RAS, cracks were seen. Nevertheless, after a few years of use, field observations on a significant roadway paved with a modified open graded frictional course asphalt mixture showed an appreciated performance. Over the course of the two-and-a-half years of observations, the pavement exhibited no signs of surface flaws or breakdown. However, a visual inspection was used to make the observations rather than a long-term performance evaluation procedure, which would have allowed for a more accurate assessment of the pavement's changing condition.

5.3.2.3.3. Construction and Demolition (C&D) Waste Materials

According to research on the use of C&D waste materials in pavement subbase applications, waste rock and RCA have geotechnical engineering qualities that are on par with or better than those of common granular subbase materials from quarries. It was also discovered that crushed brick (CB) satisfies the specifications of common quarry granular subbase materials at lower target moisture contents of 70% of the Optimum Moisture Content (OMC). To enable their use in pavement subbases, the qualities of CB, RAP, and fine recycled glass (FRG) can be further improved with additives or combined in blends with premium aggregates. Furthermore, when RCA/CB blends were evaluated using resilient modulus rather than abrasion loss, it was discovered that they could be used in base or subbase applications (Arisha et al., 2016).

5.3.2.3.4. Crumb Rubber

Both the wet process and the dry process can be used to incorporate crumb rubber into the asphalt mix, either as a bituminous material modifier or as a component of solid aggregate. Crumb rubber, which is used in asphalt surfacing applications, offers a number of benefits regardless of the processing method—wet or dry. It offers high-shear strength asphalt mixtures, which are advantageous for bearing the weight of imposed traffic and avoiding rutting in the layers of pavement underneath (Huang et al., 2007).

Higher percentages of crumb rubber modifier (CRM) in asphalt mixtures appeared to result in higher viscosity, improved resistance to rutting, and a decreased risk of low-temperature cracking. Furthermore, when compared to conventional asphalt pavement, the use of crumb rubber with WMA additive not only effectively extended the long-term performance of the pavement but also decreased the mixing and compaction temperatures of rubberized asphalt mixtures offset by crumb rubber. Regardless of rubber size or type, adding crumb rubber to the mixture containing 25% RAP helped to improve rutting resistance and increase the voids in mineral aggregate (VMA) in the Superpave mix design. Nonetheless, the modified mixtures' fatigue life was increased while their resilient modulus values decreased with an increase in rubber size, irrespective of the type of rubber (Xiao et al., 2009).

Furthermore, there are many benefits to using crumbed rubber in road construction applications, such as decreased flow value (up to 10% of crumb rubber), enhanced resistance to stripping, rutting and fatigue, and rendered temperature susceptibility. Properties like weathering and stripping resistance have been significantly improved when up to 30% of crumb rubber is added to the asphalt binder. In order to reduce noise, open-graded friction course surface layers in Arizona use crumb rubber.

5.3.2.3.5. Plastic Wastes

Plastic wastes can stabilize different pavement layers or embankments, as well as be used in asphalt mixes to coat the stone aggregate or act as a modifier of asphalt cement. In terms of stability, skid resistance, and decreased deformation, PCA (polymer-coated aggregate/plastic-coated aggregate) with up to 25% plastic waste added improved HMA performance. Because of their high stability, waste HDPE-modified binders offered superior resistance against long-term deformations. When polyethylene terephthalate from plastic bottles was used to test mixture properties in Nigeria, notable improvements were seen for 60/70 penetration grade bitumen. To stabilize embankments, recycled plastic pins (RPPs) made from plastic waste have been used (Khan et al., 2015).

5.3.2.3.6. Foundry Sands

A by-product of the ferrous and nonferrous metal casting industry, foundry sand is categorized as a fine-grained material made up of superior silica with consistent size and notable cleanliness. According to published research, sand is recycled at foundries until it may no longer be feasible to do so, at which point it is disposed of as waste sand. When thinking about it for road construction, this is the time. This material has been used as flowable fill in HMA, subbase, embankments, and backfill. Because of its high clay content, it cannot be grown in areas with poor drainage or high-water levels. Because nonferrous foundries contain a lot of heavy metallic substances, it is not advised for HMA. The majority of molding sand in Ontario, Canada, is recycled back to foundry factories, and due to concerns about moisture-induced damage, its incorporation into asphalt mixtures is only permitted up to a maximum of 15% of the total. However, given the reported technical and environmental consequences, it is no longer advised to use this material in base/subbase applications. For the foundry-modified subbase, a cement rate of 5.5% was added, and the outcomes were positive and satisfying. Additional research revealed that adding 5–15% base-subbase did not change the MDD or OMC. Nevertheless, additional research on the performance of modified base-subbase materials was recommended because the resilient modulus results were significantly lower than those of the referenced base mixtures. Even though the foamed asphalt mixture with foundry has less improved strength, stiffness, and rutting resistance than bottom ash, foundry sand was found to have high potential characteristics to substitute the natural aggregates in the CIPR mixture.

5.3.2.3.7. Coal Combustion Products/Supplementary Cementitious Material

This comprises oil-contaminated soil, fly ash, bottom ash/boiler slag, pond ash, oil sand, oil shale ash, and brake shoe fines. Fly ash is a viable mineral filler for HMA because of its ultra-tiny particles and moderate rate of heavy metals. It may be of type E, which has a lower calcium content, or type C, which has a higher calcium content. The properties of the subgrade were enhanced by the addition of 2–12% fly ash. Around 10% of type C was utilized with RAP material in the base material, and a correlation was observed between the CBR and resiliency value of the lab and field results. For subbase and base course material in light traffic pavements, it would be possible to add 40% fly ash to the RAP-modified mixture with 20% new aggregates. When compared to 1–3% of lime and cement in CIPR mixtures, fly ash can be used at a rate of 8–12%, resulting in a 25% savings. To maintain the necessary compressive strength in compliance, a rate of 25% cement and 12% cement should be maintained in the cement-stabilized base. In comparison to superior quality aggregates mixed with polymer-modified PG binder, it was discovered that inferior quality aggregates mixed with regular PG binder developed a significant moisture-damage resistance.

One benefit of bottom ash/boiler slag is that its angular, glassy surface provides a high resistance to skidding. Furnace bottom ash is the term used to describe ash extracted from blast furnaces. According to published research, fly ash is favored over bottom ash when building roads. However, in terms of the immediate boost in resilience, bottom ash usage outweighed fly ash usage. Each year, 3,000 tons of fly ash and 30,000 tons of bottom ash are produced in Brampton and Mississauga. In HMA, only 10% may be used. Vitrification has been suggested as a solution to the leaching problem that has been raised with this ash. The glassy-textured vitrified ash is what asphalt pavement should use as aggregates. Because of its water-hating nature, much less asphalt cement would be required to achieve the desired performance characteristics. The addition of bottom ash improves tensile strength and resistance to deformation; additional research is required to evaluate resistance to stripping characteristics.

Pond ash is a mixture of ash and water that is deposited in lagoons; it is distinguished by its non-crystalline texture. To improve the qualities, it is most frequently altered with additional additives like fiber and lime. The base course and subgrade layers experienced a decrease in vertical compressive strain and a decrease in horizontal tensile strain as the rate of fiber and lime mixing with the ash increased. As a result, as more fiber and lime were added, the pavement became more rigid. Once more, it makes sense to look into the impact of pond ash use on the subbase material's CBR value further.

Oil, tar, or bituminous sand is a naturally occurring deposit mixture of sand, clay, and water with a highly viscous asphalt cement that can reach 13% in Utah. It is a byproduct of the oil sand refining process used to produce crude oil. Alberta is home to the largest source of oil sand debris in the world. The US faces several challenges in the extraction of oil sand, including deposit composition, deposit size, and water availability for the extraction process. This material was originally used in Canada in 1913. The results showed poor mixing temperature and rutting resistance when used in Gap HMA, i.e., 104°C was a lower temperature than that of traditional HMA. It is necessary to conduct more thorough research on the incorporation of oil sand in HMA.

It is thought that adding ash from the extraction of oil shale can enhance the rheology of asphalt cement. Using a dynamic shear rheometer and a rotational viscometer test at a high temperature, the inclusion of 0 to 20% at increments of 5% was examined. Shear modulus rose as the rate of use increased. Phase angle results, however, showed that the modified binder's elastic behavior had stagnated. Therefore, the virgin binder's low-temperature behavior was unaffected by shale ash. It was recommended to look at alternative binder grades and conduct additional analysis on the lower temperature performance needed.

In China, brake shoe dust from the automobile industry was investigated as a filler material in HMA at percentages ranging from 0 to 7.5% at intervals of 2.5%. Dynamic modulus from t SPT showed that the modulus increased at 40°C, the highest temperature, and decreased at 5°C, the lowest temperature. The IDT indicated that the tensile strength improved only at a rate of 2.5%. A decision was reached to use 2.5% of the powdered material from brake shoes as a filler in HMA, even though resistance to deformation was not apparent.

5.3.2.3.8. Steel Slag, Blast-Furnace Slag, and Mining Wastes

Steel slag is an extremely durable, heavy, and abrasion-resistant aggregate that is especially well suited for use on surfaces that are covered in asphalt. Steel slag needs to be aged to reduce expansion when it is utilized as an aggregate in unbound base courses. When the amount of steel slag in concrete mixes increased, the strength of the mixes with lower cement contents increased, while the mixes with higher cement contents saw a decrease in strength.

Base courses, asphalt paving, and concrete have all successfully used air-cooled blast-furnace slag as an aggregate. Slag users using air cooling should stay away from stockpiles where steel slag and blast-furnace slag are mixed together. High-quality material that can be used in place of Portland cement in concrete is granulated blast-furnace slag (Collins and Ciesielski, 1994).

The majority of mining waste accumulations are located in remote areas, so even though they are frequently suitable as building materials for highways, their use has been sporadic. Mining wastes are derived from rock and soil. Using survey questionnaires across 35 US states, Austoroads (2009) examined the current state of practice regarding the economic and performance evaluation of using waste materials and by-products in highway construction or maintenance work. The DOTs of the states were surveyed to find out which pavement layer the waste materials were added to, if the cost was higher, lower, or the same as that of virgin materials, and whether the performance was superior to that of virgin materials substituted with waste products. A summary of example studies is shown in Tables 5.3 and 5.4 (El-Assaly and Ellis, 2001).

Table 5.3. Research Work Developed with the Incorporation of Steel Slag Aggregate (SSA) in Different Asphalt Mixtures

Type of Mixture	Improved Properties with SSA
HMA	Fatigue cracking resistance
HMA	Marshall stability and surface characteristics
HMA	Stiffness modulus, fatigue and rutting resistance, and indirect tensile strength
HMA	Marshall stability and temperature susceptibility
HMA	Marshall stability and fatigue cracking resistance
WMA	Marshall stability, stiffness, resilient modulus, and indirect tensile strength

WMA	Marshall stability, tensile strength, resilient modulus, moisture resistance, and rutting resistance
WMA	Indirect tensile strength, resilient modulus, and dynamic creep
Microsurfacing	Abrasion resistance, curing time, bleeding, and vertical displacement

Source: Loureiro CDA, Moura CFN, Rodrigues M, Martinho FCG, Silva HMRD, Oliveira JRM. (2022). Steel slag and recycled concrete aggregates: Replacing quarries to supply sustainable materials for the asphalt paving industry. Sustainability, 14(9):5022. https://doi.org/10.3390/su14095022.

Table 5.4. Advantages and Disadvantages of Waste Polymer Asphalt Modification

Waste Polymer	**Benefits**	**Drawbacks**
Wet Process		
PP	• Improve temperature susceptibility, rutting and load resistance, and high-temperature characteristics • Reduce cost	• Reduce resistance to fatigue and thermal cracking • Limited elastic recovery • Insoluble, phase separation, and significant crystallization
PE	• Improve aging, fracture toughness, rutting and fatigue resistance, and high-temperature characteristics • Reduce cost	• Reduce low-temperature performance • Limited elastic recovery • Phase separation and hard dispersion
EVA	• Improve temperature susceptibility; rutting, low-temperature, and fatigue cracking resistance; and storage stability	• Reduce low-temperature performance • Limited elastic recovery • Reduce resistance to moisture damage

PVC	• Improve aging, storage stability, rutting and fatigue resistance, indirect tensile strength, bonding between aggregate and asphalt, and storage stability	• High melting point • Reduce low-temperature performance
CRM	• Improve elasticity, stiffness, moisture damage, rutting and fatigue resistance, and high-temperature characteristics • Reduce cost	• High time for digestion • Phase segregation
Dry Process		
PE	• ·Improve stiffness, tensile strength, rutting and fatigue resistance, and Marshall stability	• Increase air voids • Limited workability • Reduce resistance to unaged fatigue and moisture damage
Recycled polystyrene	• Improve stiffness and rutting resistance.	• Increase air voids • resistance to fatigue
PET	• Improve stiffness, compressive strength, moisture damage and weathering resistance, and Marshall stability	• Increase air voids • Reduce resistance to rutting
CR	• Improve compressive strength, weathering resistance, and Marshall stability	• Increase air voids
PP	• Improve stiffness, rutting and fatigue resistance	• Reduce tensile strength, Marshall stability, and quotient • Reduce resistance moisture damage and skid • Reduce fracture energy

Source: Jwaida Z, Dulaimi A, Mydin MAO, Özkılıç YO, Jaya RP, Ameen A. (2023). The use of waste polymers in asphalt mixtures: Bibliometric analysis and systematic review. Journal of Composites Science, 7(10):415. https://doi.org/10.3390/jcs7100415.

5.4. SPECIFICATIONS OF PAVEMENT-RECYCLED MATERIALS

The conventional specifications advised by road authorities for virgin aggregates are sometimes not met by the properties of recycled materials in terms of gradation, strength/stiffness, plasticity, and durability/toughness/abrasion. However, guidelines exist for the use of some recycled materials, such as RAP and RCA, in road construction. These guidelines primarily draw from experiences producing aggregates from natural rocks and current guidelines for virgin aggregates. The usage of recycled materials for pavement, such as chemicals, serious contaminations, and rust stains, is subject to certain restrictions stated by road authorities.

Gradation-based testing or performance-based testing are required according to the specifications of pavement-recycled materials. It is inexcusable to fully consider performance-based specifications alone; instead, extensive laboratory measurements and a long-term field assessment are needed. Based on an overview of the South Australian requirements for RCA, a significant instance of recycled material utilized in pavement building, the stiffness is based on resilient modulus and permanent deformation ascertained by repeated loading triaxial testing (RLTT), while strength limits are based on shear strength parameters rather than California bearing ratio (CBR). These tests call for highly sophisticated equipment, which is costly and typically unavailable in standard laboratories. The only road authority that has specified performance-based laboratory testing by RLTT is the South Australian Road Authority; however, certain road authorities, i.e., Sweden mandated that the back-calculated modulus be determined through performance-based field testing using a falling weight deflectometer (Austoroads, 2009) (Table 5.5; Figure 5.8).

Table 5.5. Advantages/Benefits and Drawbacks of Construction and Demolition Waste (CDW) Recycling

Advantages	Benefits
It prevents the depletion of natural resources	Protection of natural habitats
Minimizing dependency on raw materials	Minimization of consumption of natural resources Prioritize ready-to-use materials
Reduction in necessary financial resources	Reduction in energy costs used in the extraction, processing, and transport of natural materials

Elimination of waste deposits	Minimization of greenhouse gas emissions Reductions in water and air pollution
Minimization of disposal expenses	Decrease in transport and disposal costs
Environmental protection	Contribution to climate change mitigation
Drawbacks	
Heterogeneity in the CDW composition. Difficulties in pre-screening CDWs. It may have contamination. Uncertainties about the standards to be met. Processing and crushing equipment may not be suitable. Lack of incentives from some public entities.	

Source: Loureiro CDA, Moura CFN, Rodrigues M, Martinho FCG, Silva MRD, Oliveira JRM. (2022). Steel slag and recycled concrete aggregates: replacing quarries to supply sustainable materials for the asphalt paving industry. Sustainability, 14(9):5022. https://doi.org/10.3390/su14095022.

Figure 5.8. *Recycled concrete aggregate (RCA) resulting from construction and demolition waste (CDW) processing.*

Source: Loureiro CDA, Moura CFN, Rodrigues M, Martinho FCG, Silva HMRD, Oliveira JRM. (2022). Steel slag and recycled concrete aggregates: Replacing quarries to supply sustainable materials for the asphalt paving industry. Sustainability, 14(9):5022. https://doi.org/10.3390/su14095022.

5.5. BENEFITS OF USING RECYCLED MATERIALS

Three primary categories can be used to classify the advantages of using recycled materials as alternative eco-materials: performance, economic, and environmental benefits.

1. To conserve natural resources and subsequently lower energy consumption and carbon emissions, the environmental benefits come from blending waste materials with virgin materials or from substituting a portion of the virgin materials in pavement layers with recycled materials. Additionally, the use of such materials protects people's health from potential toxicity and protects the environment from pollutants leaching into ground water and land. For instance, the majority of road authorities limit the use of recycled materials, such as asbestos, in construction projects.
2. Utilizing less expensive materials (recycled materials/by-products) lowers construction costs and yields economic benefits. Recycling also prevents the need for new or expanded landfills and lowers the cost of material transportation and disposal. The economic benefits of including the crumb rubber tires in the binders were measured in a recent study, which revealed significant material cost savings.
3. Lastly, when recycled materials are substituted for virgin aggregate or binder additives, pavement performance can occasionally be improved. When RCA materials are used instead of virgin materials, the unbound base and subbase layers' resilient modulus can be increased, extending the pavement's service life. Rubber tires can lessen fatigue cracking and improve the properties of the asphalt mix.

5.6. BARRIERS AND LIMITATIONS OF RECYCLED MATERIALS/BYPRODUCTS IN PAVEMENT CONSTRUCTION

The use of recycled materials and byproducts in pavement construction is constrained by many factors. Sorting certain waste products, like RAP or C&D demolition waste products, is a necessary first step. Manpower, site space, waste handling and sorting equipment, and management plans are all necessary for sorting. Crushing is also necessary for C&D waste materials to reach a particular gradation. Other obstacles include a lack of specifications, a lack of experience with such materials, and a shortage of skilled technicians and employees. Contractors and other stakeholders lack the expertise and/or confidence to recycle waste or by-products in the construction of pavement in many nations. Regarding the performance of these materials in the field, they require trustworthy data. Actually, there aren't many or any data available at this time regarding the economics and/or performance of the majority of by-products in the road construction industry. The agencies and contractors are always concerned about the variability in the properties of the waste materials and by-products. Another major obstacle is the absence of necessary modifications to conventional empirical relationships when waste materials are included. Additionally, extensive research programs and trial field sections are needed in order to use more recycled materials in the various pavement layers. Lastly, waste additives are resisted by contractors and highway departments/agencies because they view them as an additional material to worry about. These are the main obstacles preventing recycled and by-product materials from being used in the construction sector generally, and pavement construction specifically.

5.7. SUCCESSFUL IMPLEMENTATION OF WASTE MATERIALS IN PAVEMENT CONSTRUCTION

This section outlines current methods and tactics for getting around some of the obstacles and restrictions that prevent the successful use of waste and by-product materials in pavement construction. Stakeholder commitment is crucial to overcoming these obstacles, whether they are nontechnical or technical. Using an efficient waste management strategy is the key to successfully implementing waste and by-products in the pavement construction sector. The pavement construction industry is not environmentally friendly. Thus, the successful implementation of waste materials and by-products in the pavement construction industry sector requires the imposition of waste management plans, the execution of awareness campaigns for consultants, engineers, and government officials, worker and technician training, the development of new specifications for such materials, and the monitoring of these materials' long-term performance. To determine the ideal percentage of recycled materials and encourage the use of the materials, both long- and short-term research projects utilizing extensive laboratory testing and field trial sections are required. It is crucial for consultants and contractors to share their experiences working with these materials and to continue their education.

Encouraging the reuse of such materials requires the imposition of laws for environmental protection. For example, to reduce the amount of waste that must be disposed of in landfills, Massachusetts has implemented landfill bans for certain types of C&D waste. Bricks, concrete, wood, clean gypsum board, metals, and asphalt pavement materials were all forbidden from being dumped by the Massachusetts Department of Environmental Protection (DEP) in 2011. Governments and local governments all over the world are supporting laws that recycle C&D waste. To boost the rate of recycling C&D waste, various legislative proposals were put forth by the provincial governments of Canada.

The estimation of the quantities available from these materials for effective planning of the pavement construction projects is another crucial and significant issue for the successful use of waste/by-products. For instance, there are several models and approaches available to estimate the amounts of C&D waste.

5.8. SUSTAINABLE WASTE MANAGEMENT FOR PAVEMENT MATERIALS

Three primary categories comprise the waste hierarchy applied to pavement materials: minimize and prevent waste production, reuse and recycle, and optimize disposal. Reducing waste production is the most preferred option, while landfilling is the least preferred option. The following are the main steps in the hierarchy framework of pavement material waste management, as shown in the Figure 5.9:

- **Reduce**: By reducing the amount of waste generated at the source, reduction seeks to minimize waste production. The economical use of waste during pavement design, road construction, and pavement maintenance is known as waste reduction.
- **Reuse**: Reusing RAP, for example, from existing pavements minimizes the cost of disposal.
- **Recycle**: The most money can be saved by recycling the materials from the pavement's upper layers into new products because these are the highest-quality and priciest materials. However, the supply of recyclable waste, consumer demand for recycled goods, and the accessibility of recycling equipment all play a role in this. For instance, using RAP materials for in-place recycling to create cold asphalt mixes reduces costs and uses less energy.
- **Treat**: Treatment is the process of removing risks and contaminants before disposing of waste that cannot be recycled or reused. Sorting and separating the material from the waste is necessary for the treatment.
- **Dispose**: The least desirable option in the waste hierarchy for pavement materials is to dump them in landfills, but it is an environmentally sound choice in situations where there are no other options.

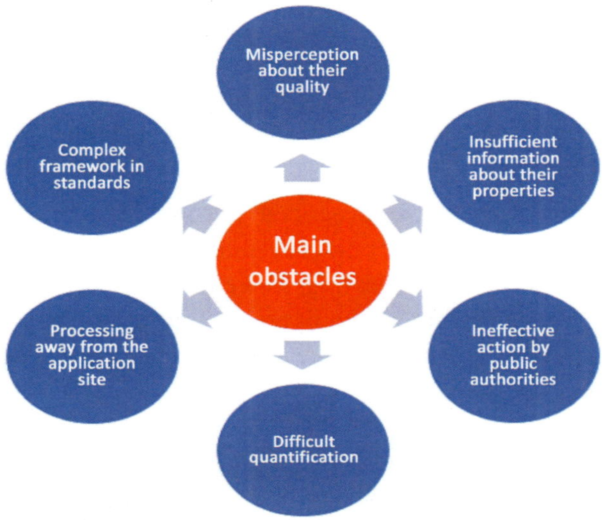

Figure 5.9. *Waste from CDW faces these issues.*

Source: Loureiro CDA, Moura CFN, Rodrigues M, Martinho FCG, Silva HMRD, Oliveira JRM. (2022). Steel slag and recycled concrete aggregates: Replacing quarries to supply sustainable materials for the asphalt paving industry. Sustainability, 14(9):5022. https://doi.org/10.3390/su14095022.

Recycled materials have the potential to produce new products that are both cost- and quality-competitive with natural materials. To cut down on pollution and save transportation costs, RCA can be produced in either stationary or portable crushers. The strategy's three primary phases are the consultative process, performance and product suitability, and knowledge transfer, as the Figure 5.10 makes clear.

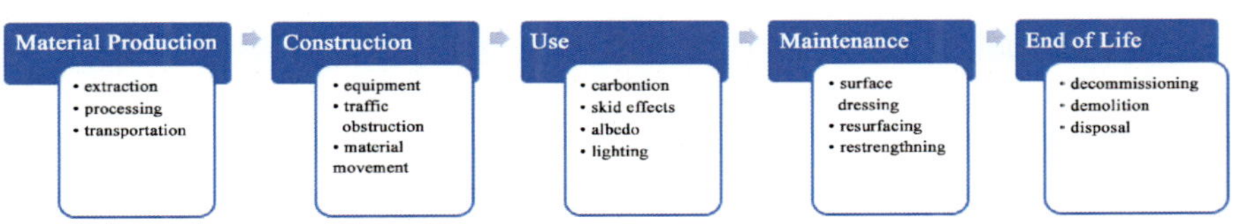

Figure 5.10. *Life cycle phases for road pavements and factors/processes.*

Source: Loureiro CDA, Moura CFN, Rodrigues M, Martinho FCG, Silva HMRD, Oliveira JRM. (2022). Steel slag and recycled concrete aggregates: Replacing quarries to supply sustainable materials for the asphalt paving industry. Sustainability, 14(9):5022. https://doi.org/10.3390/su14095022.

Customers, the industry, and stakeholders should be consulted during the consultative process stage in order to identify products, gaps in industry practices, and advantages or disadvantages of the product. The second phase involves characterizing the product by understanding its attributes and conducting performance-based testing, which involves both lab and field trials. Finally, the outcomes of the project are disseminated to public and market.

An illustration of the above-described strategy for recycling is provided in Figure 5.11. The RCA production process consists of waste collection and delivery, manufacturing and processing, as well as transport and distribution to the final destination, as shown in the figure. A quarry produces 7.5 kg of CO_2 per ton, whereas a production unit processing recycled materials produces 3 kg (Stroup-Gardiner, 2013).

Figure 5.11. *Production scheme.*

Source: Jamieson S, White G, Verstraten L. (2024). Principles for incorporating recycled materials into airport pavement construction for more sustainable airport pavements. Sustainability, 16(17):7586. https://doi.org/10.3390/su16177586.

REFERENCES

1. Abbas, A. R., Mannan, U. A., & Dessouky, S. (2013). Effect of recycled asphalt shingles on physical and chemical properties of virgin asphalt binders. *Construction and Building Materials, 45*, 162–172.
2. Agyeman, S., & Ampadu, S. I. (2015). Exploring the techno-economic feasibility of mine rock waste utilization in road works: The case of a mining deposit in Ghana. *Waste Management & Research, 34(2), 156–164.*
3. Aiban, S. A. (2005). Utilization of steel slag aggregate for road bases. *Journal of Testing and Evaluation, 34*(1), 1–11.
4. Akbulut, H., & Gürer, C. (2007). Use of aggregates produced from marble quarry waste in asphalt pavements. *Building and Environment, 42*(5), 1921–1930.
5. Al-Qadi, I. L., Elseifi, M., & Carpenter, S. H. (2007). *Reclaimed Asphalt Pavement—a Literature Review* (Review No. FHWAICT-07-001).
6. Alvergue, A., Elseifi, M., Mohammad, L. N., Cooper Jr, S. B., & Cooper III, S. (2014). Laboratory evaluation of asphalt mixtures with reclaimed asphalt shingle prepared using the wet process. *Road Materials and Pavement Design, 15*(1), 62–77.
7. Andrews, B., Sharp, K., & Robinson, P. (2009). Report on construction and demolition waste recycled materials workshop. Draft report, Contract No. RC74540-1, Supported by Zero Waste SA, Resourceco and Adelaide Resource Recovery (ARR), Australian Road Research Board (ARRB), Australia.
8. Angelone, S., Casaux, M. C., Borghi, M., & Martinez, F. O. (2017). Green pavements: Reuse of plastic waste in asphalt mixtures. *Materials and Structures*, 1–11.
9. Arisha, A. (2017). Evaluation of recycled clay masonry blends in pavement construction. *MSc. thesis*, Public Works Engineering Department, Mansoura University, Egypt.
10. Arisha, A., Gabr, A., El-Badawy, S., & Shwally, S. (2016). Using blends of construction & demolition waste materials and recycled clay masonry brick in pavement. *Proc Eng, 143*, 1317–1324.
11. Arshad, A. K., Mohammad, M., Shaffie, E., Hashim, W., & Halim, A. A. (2017). Volumetric analysis and performance of hot mix asphalt with variable RAP content. *MATEC Web Conf., 103*, 09004.
12. Arulrajah, A., Piratheepan, J., Disfani, M., & Bo, M. W. (2012). Geotechnical and geoenvironmental properties of recycled construction

and demolition materials in pavement subbase applications. *Journal of Materials in Civil Engineering, 25*(8), 1077–1088.

13. Asi, I. M., Qasrawi, H. Y., & Shalabi, F. I. (2007). Use of steel slag aggregate in asphalt concrete mixes. *Canadian Journal of Civil Engineering, 34*(8), 902–911.
14. Athanasopoulou, A., & Kollaros, G. (2015). Fly ash exploited in pavement layers in environmentally friendly ways. *Toxicological & Environmental Chemistry, 97*(1), 43–50.
15. Austroads. (2009). *Guide to Pavement Technology: Part 4E: Recycled Materials* (Publication No. AGPT04E/09). Project No. TP1565, Austroads, Melbourne, Australia.
16. Aziz, M. M. A., Rahman, M. T., Hainin, M. R., & Bakar, W. A. W. A. (2015). An overview on alternative binders for flexible pavement. *Construction and Building Materials, 84*, 315–319.
17. Bakis, R., Koyuncu, H., & Demirbas, A. (2006). An investigation of waste foundry sand in asphalt concrete mixtures. *Waste Management & Research, 24*(3), 269–274.
18. Barišić, I., Dimter, S., & Rukavina, T. (2014). Strength properties of steel slag stabilized mixes. *Composites Part B: Engineering, 58*, 386–391.
19. Bassani, M., Santagata, E., Baglieri, O., Ferraris, M., Salvo, M., & Ventrella, A. (2009). Use of vitrified bottom ashes of municipal solid waste incinerators in bituminous mixtures in substitution of natural sands. *Advances in Applied Ceramics, 108*(1), 33–43.
20. Biligiri, K. P., & Kaloush, K. E. (2010). Laboratory evaluation of noise-dampening properties of asphalt mixtures. In *39th International Congress on Noise Control Engineering, INTER-NOISE 2010* (Vol. 9, pp. 7298–7307), Lisbon, Portugal.
21. Carlson, D. D., Way, G. B., Zareh, A., Kaloush, K. E., & Biligiri, K. P. (2009). Noise characteristics and field performance of five different wearing courses in Arizona. In *Asphalt Rubber 2009 Conference Proceedings*, 2–4 November, Nanjing, China.
22. Cascione, A. A., Williams, R. C., & Yu, J. (2015). Performance testing of asphalt pavements with recycled asphalt shingles from multiple field trials. *Construction and Building Materials, 101*, 628–642.
23. Chandra, S., Kumar, P., & Feyissa, B. A. (2002). Use of marble dust in road construction. *Road Materials and Pavement Design, 3*(3), 317–330.
24. Chesner, W. H., Collins, R. J., & MacKay, M. (1998). *User Guidelines for*

Waste and by-Product Materials in Pavement Construction (Publication No. FHWA-RD-97-148). FHWA, Washington, DC.

25. Choudhary, A. K., Jha, J. N., & Gill, K. S. (2010). Utilization of plastic wastes for improving the sub-grades in flexible pavements. *Paving Materials and Pavement Analysis*, 320–326.
26. Collins, R. J., & Ciesielski, S. K. (1994). *Recycling and Use of Waste Materials and by-Products in Highway Construction* (Synthesis of Highway Practice Series 199). National Research Council (U.S.), Transportation Research Board National Cooperative Highway Research Program, Washington, D.C.: National Academy Press.
27. Copeland, A. (2011). *Reclaimed Asphalt Pavement in Asphalt Mixtures: State of the Practice*. Federal Highway Administration, Office of Research, Development and Technology, Turner-Fairbank Highway Research Center, McLean.
28. de Rezende, L. R., de Carvalho, J. C., & Palmeira, E. M. (2005). The use of alternative and improved construction materials and geosynthetics in pavements. *Elsevier Geo-Engineering Book Series, 3*, 765–786.
29. de Rezende, L. R., Marques, M. O., & da Cunha, N. L. (2015). The use of non-conventional materials in asphalt pavement base. *Road Materials and Pavement Design, 16*(4), 799–814.
30. Dimter, S., Rukavina, T., & Dragèevic, V. (2011). Strength properties of fly ash stabilized mixes. *Road Materials and Pavement Design, 12*(3), 687–697.
31. Dumitru, I., & Walter, P. (2001). The use of recycled construction materials in the manufacture of road building materials. In *Beneficial Use of Recycled Materials in Transportation Applications* (pp. 13–15, Road Transport Authority, Arlington, Australia).
32. El-Assaly, A., & Ellis, R. (2001). Evaluation of recycling waste materials and by-products in highway construction. *The International Journal of Sustainable Development & World Ecology, 8*(4), 299–308.
33. Emersleben, A., & Meyer, N. (2012). The use of recycled glass for the construction of pavements. In *GeoCongress 2012: State of the Art and Practice in Geotechnical Engineering* (pp. 1642–1649).
34. Fall, M., & Samb, S. (2008). Pore structure of cemented tailings materials under natural or accidental thermal loads. *Journal of Materials Characterization, 59*(5), 598–605.
35. Fall, M., Belem, T., Samb, S., & Benzaazoua, M. (2007). Experimental characterization of the stress-strain behavior of cemented paste backfill. *Journal of Materials Science, 42*, 3914–3922.

36. Fall, M., Célestin, J., Pokharel, M., & Touré, M. (2010). A contribution to understanding the effects of curing temperature on the mechanical properties of mine cemented tailings backfill. *Engineering Geology, 114*(3), 397–413.
37. Fall, M., Célestin, J., Pokharel, M., & Touré, M. (2010). A contribution to understanding the effects of temperature on the mechanical properties of cemented mine backfill. *Engineering Geology, 14*(3, 4), 397–413.
38. Gabr, A., Cameron, D., Andrews, R., & Mitchell, P. (2011). Comparison of specifications for recycled concrete aggregate for pavement construction. *Journal of ASTM International, 8*(10), 1–15.
39. Guney, Y., Aydilek, A. H., & Demirkan, M. M. (2006). Geoenvironmental behavior of foundry sand amended mixtures for highway subbases. *Waste Management, 26*(9), 932–945.
40. Hesp, S. (2015). Interview with Lisa Laco. *Superior Morning with Lisa Laco*, CBC.
41. Hınıslıoğlu, S., & Ağar, E. (2004). Use of waste high density polyethylene as bitumen modifier in asphalt concrete mix. *Materials Letters, 58*(3), 267–271.
42. Holbert, C., Drelich, J., Zmierczak, W., & Miller, J. D. (1997). Viscosity of bitumen-crumb rubber blend (new paving material). *Petroleum Science and Technology, 15*(5, 6), 523–543.
43. Huang, Y., Bird, R. N., & Heidrich, O. (2007). A review of the use of recycled solid waste materials in asphalt pavements. *Resources, Conservation and Recycling, 52*(1), 58–73.
44. Kalantar, Z. N., Karim, M. R., & Mahrez, A. (2012). A review of using waste and virgin polymer in pavement. *Construction and Building Materials, 33*, 55–62.
45. Kandhal, P. S., & Mallick, R. B. (1998). *FHWA, Pavement Recycling Guidelines for State and Local Governments* (Report No. FHWA-SA-98-042). Federal Highway Administration, Office of Engineering/ Office of Technology Applications, Washington, DC.
46. Kang, D. H., Gupta, S. C., Ranaivoson, A. Z., Siekmeier, J., & Roberson, R. (2011). Recycled materials as substitutes for virgin aggregates in road construction: I. Hydraulic and mechanical characteristics. *Soil Science Society of America Journal, 75*(4), 1265–1275.
47. Karmakar, S., & Roy, T. K. (2016). Effect of waste plastic and waste tires ash on mechanical behavior of bitumen. *Journal of Materials in Civil Engineering, 28*(04016006).

48. Khan, M. S., Hossain, S., & Kibria, G. (2015). Slope stabilization using recycled plastic pins. *Journal of Performance of Constructed Facilities, 30*(3), 04015054.

49. Khan, R. A., & Ganesh, A. (2016). The effect of coal bottom ash (CBA) on mechanical and durability characteristics of concrete. *Journal of Building Materials and Structures, 3*(1), 31–42.

50. Khosla, N. P., Nair, H., Visintine, B., & Malpass, G. (2012). Effect of reclaimed asphalt and virgin binder on rheological properties of binder blends. *International Journal of Pavement Research and Technology, 5*(5), 317.

51. Kim, Y. R., Pinto, I., & Park, S. W. (2012). Experimental evaluation of anti-stripping additives in bituminous mixtures through multiple scale laboratory test results. *Construction and Building Materials, 29*, 386–393.

52. Krayushkina, K., Prentkovskis, O., Bieliatynskyi, A., & Junevičius, R. (2012). Use of steel slags in automobile road construction. *Transport, 27*(2), 129–137.

53. Lav, M. A., & Lav, A. H. (2014). Effects of stabilization on resilient characteristics of fly ash as pavement material. *Construction and Building Materials, 54*, 10–16.

54. Lee, J., Edil, T. B., Benson, C. H., & Tinjum, J. M. (2013). Building environmentally and economically sustainable transportation infrastructure: Green highway rating system. *Journal of Construction Engineering and Management, 139*(12), A4013006.

55. Lee, J., Edil, T., Tinjum, J., & Benson, C. (2010). Quantitative assessment of environmental and economic benefits of recycled materials in highway construction. *Transportation Research Record, 2158*, 138–142.

56. Lee, S. J., Akisetty, C. K., & Amirkhanian, S. N. (2008). The effect of crumb rubber modifier (CRM) on the performance properties of rubberized binders in HMA pavements. *Construction and Building Materials, 22*(7), 1368–1376.

57. Lee, S. W., & Fishman, K. L. (1993). Waste products as highway materials in flexible pavement system. *Journal of Transportation Engineering, 119*(3), 433–449.

58. Li, L., Benson, C. H., Edil, T. B., Hatipoglu, B., & Tastan, O. (2007). Evaluation of recycled asphalt pavement material stabilized with fly ash. In *Soil and Material Inputs for Mechanistic-Empirical Pavement Design* (pp. 1–10).

59. Maher, M. H., & Popp, W. J., Jr. (1997). Recycled asphalt pavement as a

base and subbase material. In M. A. Wasemiller & K. B. Hoddinott (Eds.), *Testing Soil Mixed with Waste or Recycled Materials* (ASTM STP 1275, pp. 42–53). ASTM International.

60. Mashaan, N. S., & Karim, M. R. (2014). Waste tyre rubber in asphalt pavement modification. *Materials Research Innovations, 18*(S6), S6–6.

61. Modarres, A., & Ayar, P. (2016). Comparing the mechanical properties of cold recycled mixture containing coal waste additive and ordinary Portland cement. *International Journal of Pavement Engineering, 17*(3), 211–224.

62. Moon, K. H., Falchetto, A. C., Marasteanu, M., & Turos, M. (2014). Using recycled asphalt materials as an alternative material in asphalt pavements. *KSCE Journal of Civil Engineering, 18*(1), 149–159.

63. Moon, K. H., Falchetto, A. C., Marasteanu, M., & Turos, M. (2014). Using recycled asphalt materials as an alternative material source in asphalt pavements. *KSCE Journal of Civil Engineering, 18*(1), 149.

64. Mousa, E., Azam, A., El-Shabrawy, M., & El-Badawy, S. (2017). Laboratory characterization of reclaimed asphalt pavement for road construction in Egypt. *Canadian Journal of Civil Engineering, 44*(999), 417–425.

65. Nasir, O., & Fall, M. (2008). Shear behavior of cemented paste-fill-rock interfaces. *Engineering Geology, 101*(3, 4), 146–153.

66. Paranavithana, S., & Mohajerani, A. (2006). Effects of recycled concrete aggregates on properties of asphalt concrete. *Resources, Conservation and Recycling, 48*(1), 1–12.

67. Pasetto, M., & Baldo, N. (2012). Laboratory investigation on foamed bitumen bound mixtures made with steel slag, foundry sand, bottom ash, and reclaimed asphalt pavement. *Road Materials and Pavement Design, 13*(4), 691–712.

68. Pettinari, M., & Simone, A. (2015). Effect of crumb rubber gradation on a rubberized cold recycled mixture for road pavements. *Materials & Design, 85*, 598–606.

69. Rao, A., Jha, K. N., & Misra, S. (2007). Use of aggregates from recycled construction and demolition waste in concrete. *Resources, Conservation and Recycling, 50*(1), 71–81.

70. Reid, J. M. (2000). The use of alternative materials in road construction. In *International Symposium on Unbound Aggregates in Roads–UNBAR* (Vol. 5).

71. Reid, M. (1998). ALT-MAT: Alternative materials in road construction.

In *Polluted+ Marginal Land-98: Proceedings of the 5th International Conference on Re-Use of Contaminated Land and Landfills* (7–9 July, 1998). Brunel University, London.

72. Ribeiro de Rezende, L., da Silveira, L. R., de Araújo, W. L., & da Luz, M. P. (2013). Reuse of fine quarry wastes in pavement: Case study in Brazil. *Journal of Materials in Civil Engineering, 26*(8), 05014003.
73. Saghafi, B., Nageim, H. A., VisuliosMPhil, P., & Ghazireh, N. (2012). Use of waste limestone dust and steel slag in UK highways type 1 unbound mixtures. *Proceedings of the Institution of Civil Engineers-Construction Materials, 166*(2), 99–107.
74. Saltan, M., & Fındık, F. S. (2008). Stabilization of subbase layer materials with waste pumice in flexible pavement. *Building and Environment, 43*(4), 415–421.
75. Saride, S., Avirneni, D., Javvadi, S. C. P., Puppala, A. J., & Hoyos, L. R. (2015). Evaluation of fly ash treated reclaimed asphalt pavement for base/ subbase applications. *Indian Geotechnical Journal, 45*(4), 401–411.
76. Saride, S., Puppala, A. J., & Williammee, R. (2010). Civil Engineers Ground Improvement 163 February 2010 Issue GI1. *sta, 717*(715).
77. Sojobi, A. O., Nwobodo, S. E., & Aladegboye, O. J. (2016). Recycling of polyethylene terephthalate (PET) plastic bottle wastes in bituminous asphaltic concrete. *Cogent Engineering*.
78. Sonnevera International Corporation. (2006). *Construction, Renovation and Demolition Waste Materials: Opportunities for Waste Reduction and Diversion* (Final Report, Pub. No: J/105). Environmental Partnerships and Education Branch, Alberta Environment.
79. Souliman, M., El-Hakim, R. A., Davis, M., & Walubita, L. (2017). Quantifying the mechanistic and economic impacts of using asphalt rubber mixtures. In L. Struble, G. Tebaldi (Eds.), *Materials for Sustainable Infrastructures: Proceedings of the 1st GeoMEast International Congress and Exhibition, Egypt 2017* (pp. 1–11). Springer.
80. Stroup-Gardiner, M. (2013). *Recycled Materials and Byproducts in Highway Applications* (Vol. 1). Transportation Research Board.
81. Tuncan, M., Tuncan, A., & Cetin, A. (2003). The use of waste materials in asphalt concrete mixtures. *Waste Management & Research, 21*(2), 83–92.
82. Vasudevan, R., Sekar, A. R. C., Sundarakannan, B., & Velkennedy, R. (2012). A technique to dispose waste plastics in an ecofriendly way—Application in construction of flexible pavements. *Construction and Building Materials, 28*(1), 311–320.

83. Wang, Y., Wang, G., & Ahn, Y. H. (2014). Effects of using recycled asphalt shingles in open-graded friction course on flexible pavement performance. *Journal of Testing and Evaluation, 42*(5), 1–10.

84. Way, G. B., Carlson, D. D., Sousa, J. B., Kaloush, K. E., & Biligiri, K. P. (2010). Introduction to asphalt-rubber pavement noise reducing characteristics. In *39th International Congress on Noise Control Engineering, INTER-NOISE 2010* (Vol. 9, pp. 7408–7417). Lisbon, Portugal.

85. Wu, D., Fall, M., & Cai, S. J. (2013). Coupling temperature, cement hydration, and rheological behavior of cemented paste backfill structures. *Minerals Engineering, 42*, 76–87.

86. Xiao, F., Amirkhanian, S. N., Shen, J., & Putman, B. (2009). Influences of crumb rubber size and type on reclaimed asphalt pavement (RAP) mixtures. *Construction and Building Materials, 23*(2), 1028–1034.

87. Xiao, F., Zhao, P. W., & Amirkhanian, S. N. (2009). Fatigue behavior of rubberized asphalt concrete mixtures containing warm asphalt additives. *Construction and Building Materials, 23*(10), 3144–3151.

88. Yang, J., Ddamba, S., UL-Islam, R., Safiuddin, M., & Tighe, S. L. (2013). Investigation on use of recycled asphalt shingles in Ontario hot mix asphalt: A Canadian case study. *Canadian Journal of Civil Engineering, 41*(2), 136–143.

89. Yazoghli-Marzouk, O., Vulcano-Greullet, N., Cantegrit, L., Friteyre, L., & Jullien, A. (2014). Recycling foundry sand in road construction—field assessment. *Construction and Building Materials, 61*, 69–78.

CHAPTER 6

BIO-BASED STABILIZATION PRODUCTS FOR ROAD CONSTRUCTION MATERIALS

Contents

INTRODUCTION

Road development is understandably being pushed by the expanding infrastructure and growing urban populations of developing nations. It is critical to evaluate the possible effects of large-scale road and transportation projects given the unparalleled pace and scope of these initiatives. Building new roads is essential to developing nations' socioeconomic development. Accordingly, badly planned or executed road projects may lead to significant budget overruns, a lack of natural resources, and detrimental environmental effects (such as increased global CO_2 emissions) (Alamgir et al., 2017). Generally speaking, concrete roads will always need some maintenance and repair; ironically, though, this will only be necessary less frequently than asphalt roads. This is caused by unanticipated weathering or wear, microcrack formations, overloading by accident, increased porosity, or other phenomena associated with particular environmental conditions changing too quickly to preserve functional properties (Jonkers et al., 2016).

The most common building material used worldwide is ordinary Portland cement (OPC), which is a typical component of civil engineering construction. Cement production, however, is extremely energy-intensive and environmentally unfriendly; in 2013, it reached 10,000 million tons/year, and within the next 40 years, it is expected to increase by 100% (Pacheco-Torgal and Labrincha, 2013). Greenhouse gas (GHG) emissions from road projects typically occur during all stages of their lifecycle (i.e., building and running, and to a lesser degree, repair and upkeep) (Alzard et al., 2019). The construction industry is predicted to grow rapidly to accommodate the growing urban population, which is expected to reach 6.4 billion people by 2050 (Pacheco-Torgal, 2016). According to recent estimates, urban land cover will increase by 1.2 million km^2 by 2030. As a result, there will be an increase in demand for substitute building materials.

Building and maintaining roads is expensive in developing nations. The construction industry has made a concentrated effort to lower construction costs, especially when dealing with difficult ground conditions. Implementing environmentally friendly and sustainable road development is difficult, though. The value of South Africa's roads as an asset is estimated by the South African National Roads Agency Limited (SANRAL) to be $200 billion (2014), of which the paved road network accounts for 80% (or roughly $160 billion). A low-volume paved rural road typically costs $300,000–500,000 per kilometer to build and maintain, while a heavy freeway structure can cost millions of dollars per kilometer (United States Dollars or USD) to build and maintain (National Treasury, 2015; Siyepu, 2016).

Soil roads are built in developing nations for low traffic areas, such as estate roads or rural access roads. There is a lot of evidence in South Africa (SA) that the country's current road infrastructure is not being properly maintained, and the construction of new roads has been put on hold. The ongoing rise in traffic and the moisture damage brought on by heavy rains exacerbate this. Future development prospects are also harmed by an unreliable road infrastructure. Unproclaimed roads are an excellent example of inadequate infrastructure since they are, at most, gravel roads without all-weather access. According to Paige-Green (2008), these kinds of roads may have detrimental effects on the local communities' economic health in addition to having detrimental effects on the environment (i.e., dust clouds, soil lost to erosion).

New technologies like electro kinetics, deep mixing, nanomaterials for sensors and sensing, and the use of microorganisms for groundwater remediation are being researched by the construction industry (Wijeyesekera, 2014). By incorporating ideas and methods from other disciplines, these have significantly improved the practice of road construction and produced a "new" interdisciplinary approach to asset management in civil and geotechnical engineering. Among the more recent examples of this trend is the development of microbiological processes for improving the properties of soil.

A potentially economical and environmentally friendly engineering method to strengthen and repair the current road base, sub-base, or subgrade materials for a longer lifespan is bio-based exploration for use in construction (Pei et al., 2015). Bacillus species are examples of biological stabilizers that generate a broad spectrum of environmentally acceptable metabolites and additives that provide accumulative advantages for compaction and stabilization in structural applications (Samang et al., 2018), including financial savings, environmental advantages, and performance-based rewards. Numerous Bacillus species and their derivatives have been explored as possible substitutes for traditional chemical soil stabilizers, especially in difficult-to-reach areas. This is undoubtedly a superb illustration of how to apply multifaceted green technologies to support interdisciplinary research and cooperation in the fields of geomicrobiology, biotechnology, green chemistry, and road/pavement engineering. It is critical to pursue these greener technologies, which calls for the creation of new environmentally friendly materials to replace conventional cement through research and development. The importance of applying geomicrobiological research to catalyze different reactions that stabilize *in-situ* soil for road applications using novel bio-enzymes, biopolymers, or waste, like fly ash. It is feasible to develop engineering solutions for transient and/or permanent geotechnical engineering problems thanks to this (including their mode(s) of action).

The development of novel and inventive, yet ecologically sound, bio-based product applications is highly financially motivated. Finding high-volume, low-cost uses for bio-based products in in-place soils and turning waste into products with value additions are the challenges. Numerous geotechnical applications, including dust suppression, dam control, wind soil erosion control, mitigating earthquake liquefaction, building reactive barriers, and long-term stabilization of contaminated soils, could be addressed by this strategy. Regarding their mode of action, this offers a chance to identify new microorganisms by appropriately evaluating their attributes, such as the secretion of one or more metabolic products, interesting enzymes, or biopolymers, and exploring them as soil additives. It becomes evident that a number of applied research topics in the biotechnology sector need to be addressed due to a lack of knowledge on bio-based soil stabilization. For example, suitable upstream procedures for the production of biomass, downstream procedures that produce stable products (including field testing and the required in-vitro mechanisms), well-formulated products, and the proper use of each soil stabilizer in a range of environmental circumstances.

6.1. CHALLENGES AND CONVENTIONAL STABILIZATION METHODS USED BY THE CONSTRUCTION INDUSTRY

The process of enhancing a soil's technical and physical characteristics in order to reach a particular performance target is known as "soil stabilization" (Latifi et al., 2016). The process of stabilizing *in-situ* material is thought to be crucial for enhancing the treated soil material's strength and performance, which helps to extend the subsurface's service life and produce a more cost-effective design.

One of the main factors in the geotechnical design criteria for site selection is the subsurface's bearing capacity. The following options are taken into consideration when the soil's bearing capacity is determined to be low: (i) modify the design to accommodate the site's conditions; (ii) remove and replace the *in-situ* soil; or (iii) abandon the site. The number of abandoned sites associated with unfavorable bearing capacities has skyrocketed, leading to a shortage of land and a rise in the demand for natural resources. The areas that are impacted include those that are liquefiable, have high levels of clay material covering them, have organic soils that can be problematic due to their high compressibility and low shear strength, are contaminated land, and are experiencing landslides (Makusa, 2013). Unusual soil can be identified by its abrupt decrease in volume (i.e., collapsible soils), following exposure to diverse environmental factors like rising humidity, wind-deposited silt or sand, or submersion in water, thereby changing the cementation bonds between soil particles (Ayeldeen et al., 2017). Lack of rainwater diversion is the root cause of drainage issues and can result in pavement surface or road structure deterioration. The primary cause of this is moisture retention in the road prism, specifically in the subgrade soils associated with common subgrade issues.

Pavement and geotechnical engineers are particularly concerned about the negative effects of rising moisture content on soil behavior. When the moisture content is at or below the optimum (dry-side of optimum), soil performs exceptionally well; but, when the moisture content rises above the optimum (wet-side of optimum), soil strength decreases noticeably. In areas with higher rainfall, sensitive soils have a tendency to swell as moisture content increases. This causes the subsurface soil to become too weak to support the pavement or road loads, which can lead to potholes and collapsed road surfaces. Weak *in-situ* soil replacement is not always a good option, particularly in road or pavement developments because of the ongoing construction cost dilemma in financially strapped nations, which is the usual situation in the majority of developing nations.

There's a chance to make unpaved roads in rural and urban areas better. Roads without a binder or chemical additive that have a surface coarse of unbound aggregate (gravel) are referred to as unpaved roads. When it comes to unpaved or earth roads, good wearing coarse should be able to withstand the abrasive action of traffic, erosion caused by wind and water, and excessive dust or slippage during dry weather. As unpaved roads are used, they gradually lose their coarse material due to both surface material erosion from heavy rainfall and traffic. Over time, the fines content decreases, leaving loose, coarse material that tends to accumulate. Roadside vegetation and fauna, agricultural productivity, road safety, and the quality of the air, soil, and water are all impacted by fugitive dust, which is created by loose, coarse material in traffic.

Furthermore, it is commonly known that exposure to high concentrations of airborne particulates over an extended period of time can lead to significant health issues caused by dust generated by traffic. Road surface degradation brought on by this material loss necessitates re-gravelling in order to restore the road's usability. Repairing the road and distributing fresh aggregates over the damaged area are all part of the re-gravelling process. Re-gravelling raises maintenance costs, making it both financially and environmentally unfeasible. Therefore, road engineers' approach is to use chemical treatment to enhance the *in-situ* materials' engineering properties. The decrease in maintenance expenses is the advantage of treated surface material (using substitute nontraditional stabilizers). This is because there is less gravel loss, which means there is less need for frequent grading and re-gravelling. Managed gravel loss lowers airborne pollutants and dust loading, both of which are harmful to human health.

Over the course of the road's initial and extended life cycles, it is evident that there are benefits and cost savings. Because ongoing initiatives will reduce traffic-generated dust from treated roads and improve safety and quality of life, they are crucial for developing solutions to the current problems in road construction.

Conventional methods for enhancing soil structure and engineering properties include geotextiles, dynamic compaction, drainage of vegetation, stabilization of soil-lime, stabilization of soil-cement, and injection of synthetic materials into the pore space to bind soil particles together, such as micro-fine cement, epoxy, acrylamide, phenoplasts, silicates, and polyurethane (Saleh et al., 2019). A range of chemicals, jetting, and permeation grouting techniques are used to achieve this. It should be highlighted that these methods raise significant environmental risks because, aside from sodium silicate, all chemical grouts have the potential to be toxic or dangerous (Bryson and El Naggar, 2013).

It is challenging to find a construction site that satisfies all design specifications in the majority of civil engineering projects without requiring ground modification. In order to satisfy the design requirements, the current practice is to alter the engineering properties of the naturally problematic soils. This results in the use of alternative stabilizers, which are non-traditional/biochemical and one of several techniques for improving soil. Practically speaking, this might result in a cement substitute, but that would all depend on price. The makeup of non-traditional additives varies, as does the way they interact with soil chemically and physically. Regretfully, little is understood about their basic stabilization mechanisms and how they interact with various geotechnical materials. Even in the absence of research, an eco-friendly solution is still beneficial because larger construction companies are willing to invest in eco-friendly building materials because they are aware of their environmental impact.

6.2. THE USE OF BIO-BASED INNOVATIONS FOR POTENTIAL ROAD APPLICATION

In place of traditional chemical stabilizers, a number of microorganisms and other bio-based products, including secondary metabolites, enzymatic materials, and polymeric materials, have been explored as viable options for the creation of sustainable road infrastructure. Therefore, for these bio-based materials and technologies to be used in construction on a larger scale, a deeper understanding of them is necessary. The use of bacteria in bio-based admixtures produced through industrial fermentation processes is becoming more popular for a variety of reasons, including the bacteria's quick rates of biosynthesis, high yields of desired products, quick growth on cheap culture media that can be purchased in large quantities, flexibility in genetic modification, and non-pathogenic characteristics.

Based on the outcomes of the microbial treatment of the soil, at least eight different types of biotechnological processes related to construction can be identified (Figure 6.1). Regarding the basic features of research methods, two biological processes stand out among these: (i) bioclogging, which is the biological production of pore-filling materials to significantly reduce the hydraulic conductivity of soil or porous matrix; and (ii) biocementation, which is the *in-situ* microbial production of particle-binding materials to increase the shear strength of soil. Further support for this comes from DeJong et al. (2010) claiming that in the field of soil stabilization, these approaches enable a comprehensive perspective and significant features. The microbes most appropriate for these methods, i.e., facultative anaerobic and microaerophilic bacteria that cause soil bio-clogging, biocementation, and bioaggregation, though anaerobic fermenting bacteria have also been proposed.

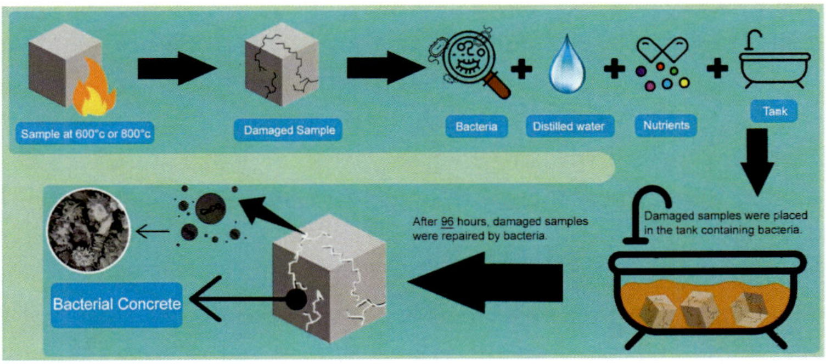

Figure 6.1. *Microbial biotechnology for cracks.*

Source: Nimafar M, Samali B, Hosseini SJ, Akhlaghi A. (2021). Use of bacteria externally for repairing cracks and improving properties of concrete exposed to high temperatures. Crystals, 11(12):1503. https://doi.org/10.3390/cryst11121503.

6.3. LARGE SCALE LABORATORY TESTING AND *IN-SITU* DEMONSTRATIONS OF BIO-BASED ADDITIVES

6.3.1. Microbial Bio-Cement

Microbial induced calcium carbonate precipitation (MICCP), also known as microbial bio-cement, is one of the most researched techniques for uniformly precipitating calcite in soils. This process involves biological action to raise pH levels and produce supersaturated conditions. For instance, the alkalophilic soil bacterium *Bacillus pasteurii* breaks down urea into carbon dioxide (CO_2) and ammonia (NH_3) by using its highly active urease enzyme. NH_3 is changed into NH_4^+ in the presence of water, and CO_2 equilibrates with carbonic acid, carbonate, and bicarbonate ions in a pH-dependent manner. The reaction with Ca^{2+} and the precipitation of calcite ($CaCO_3$) requires an alkaline environment, which is provided by an increase in pH.

Cheng et al. (2013) used the unconfined compressive strength (UCS) of MICCP-treated samples before and after freeze-thaw (FT) cycling to demonstrate bio-enzymatic mediated soil improvement. The results showed that less than 10% of the strength decreased from > 800 kPa (before FT cycle) for the OPC treated samples > 600 kPa (before FT cycle), and usually causes serious damage, as expected with ~ 40% of the strength decreasing. Shear strength was examined by Ng et al. (2012) found that the treated to untreated shear strength ratio increased from 1.40 to 2.64 using the *B. megaterium* MICCP process. A minimum compressive strength of 300 kPa was documented by van Paassen et al. (2010) revealed findings that were comparable to those of Nafisi et al. (2019) which demonstrate that the mass of carbonate and type of sand affect the tensile strength, which ranges from 210 kPa to 710 kPa. Van Paassen et al. (2010) also found that the permeability of treated soils decreased by 60% at ~100 kg/m^3 of $CaCO_3$ precipitation; however, as the solution's concentration increased, calcite crystals clogged pore spaces, reducing permeability by 50–99% when 1 M cementation solution was used. This is because the crystals that form at the contact points can preserve the connection of pores without limiting the mobility of pore water. Triaxial testing is advised to evaluate how bio-cemented soils respond to both monotonic and cyclic loadings, even though these tests characterize the strength properties on a large scale. This is because the triaxial test replicates the behavior of soil in the field intended for road applications.

Cheng et al. (2013) describes how the mechanical behavior of the bio-cemented sand in triaxial tests increased as the $CaCO_3$ content increased at all saturation levels, resulting in an increase in effective shear strength (i.e., cohesion, frictional angle). Using coarse sand, the precipitated crystals improved cohesion more effectively at a lower saturation level than increasing the frictional angle. When comparing coarse and fine sand with identical saturation conditions and $CaCO_3$ contents, the coarse sand showed a greater friction angle. The main effects of small soil particles, according to the researchers, are to: (i) increase the number of interparticle contact points for the MICCP process; and (ii) lower the stress acting at each particle contact. When cementation first starts around a particle contact, the MICCP process operates most effectively there. As cementation continues, its impact lessens. It would be more efficient to reallocate the $CaCO_3$ crystals to two contact locations (connection points) as opposed to one because this would increase the interparticle connections. The contact stress diminishes in tandem with the particle radius squared. Consequently, two compounding advantages of smaller particles are (i) higher MICCP; and (ii) reduced particle contact stresses (Cheng et al., 2013). After evaluation, scientists conclude that these important criteria allow for a better selection of the highest-performing bio-additives of interest that satisfy technical requirements like durability properties. These criteria are similar in test methods and the experimental data supports our conclusion. The information gathered here can also be utilized to inform the design of experiments for small- to large-scale field testing, taking into account the right dosage methods, the length of time needed for curing, and the amount of clay in the soil. These factors will have additional implications for the strength requirements that must be determined, such as how materials react to pressure, temperature, or moisture.

The UCS value and the $CaCO_3$ content of the treated soils have an exponential relationship, which indicates that even though the amount of precipitated $CaCO_3$ is the same, the mechanical response can differ based on the $CaCO_3$ precipitation/crystals' mechanism of action. Additionally, the microbes need to have high urease activity in order to carry out the reaction, and the mode of action (MICCP) may vary depending on the strain. Because the bacterial cell wall is negatively charged, the effect can be amplified; more cations from the surrounding environment, such as Ca^{2+}, are drawn to the bacteria and deposit on their cell surface.

Figure 6.2. *A cubic sample: (a) damaged at 600°C; and (b) repaired by type I bacteria.*

Source: Nimafar M, Samali B, Hosseini SJ, Akhlaghi A. (2021). Use of bacteria externally for repairing cracks and improving properties of concrete exposed to high temperatures. Crystals, 11(12):1503. https://doi.org/10.3390/cryst11121503.

The first step in the formation of a crystal from a solution is the nucleation site, which is the cell surface where the precipitation of $CaCO_3$ occurs as a result of the Ca^{2+} ions' subsequent reaction with the CO_3^{2-} ions. Numerous published studies reveal a number of Bacillus species possess the capacity to precipitate structurally-sound calcium carbonate; however, a restricted range of microorganisms emphasizes the triaxial and California bearing ratio (CBR) tests for possible use in pavement and roads (Figure 6.2).

MICCP is being developed and tested in the lab and on small fields for various construction and geotechnical uses. Road engineers are fundamentally concerned that MICCP could lead to crust formation, which would increase runoff and impermeability. They are also concerned about the impact of pH on plants and animals, calcite degradation from rainwater's acids, and sheet runoff, which would be especially problematic for karst areas. As a result, the field-scale evaluation of microbial or enzyme bio-cement technology for road applications under actual conditions has been rare.

When it comes to utilizing alternative materials and improving *in-situ* materials while reducing environmental impact, innovation is needed. Investigating the use of novel microorganisms and their products to meet specific performance requirements—such as longer-lasting, stronger, and more resilient road surfacing—is therefore one possible solution (smoother, better road substructure). The papers by Ghosh et al. (2005) are among those that describe biological stabilizing methods as well as Raut et al. (2014), demonstrating the numerous uses of biological agents in related fields, including slope stabilization, building settlement reduction, liquefaction and erosion prevention, impermeable and reactive barriers for groundwater protection, aquifer and energy storage, and carbon sequestration as well as dam safety. The commercialization of bio-based products is hindered by the fact that most studies on microbial geo-technology for soil road construction have large gaps in ex situ field-testing research. This offers the chance to carry out research in this area that could lead to the development of novel procedures and valuable products for this quickly expanding sector.

The application of enzymes and bacteria is used in efforts to improve soil. Alkaliphillic heterotrophic spore-forming bacteria seem like good options for enhancing concrete's durability because they don't produce any other potentially harmful metabolic byproducts, like ammoniums from the ureolytic pathway. Furthermore, alkaliphillic species of Bacillus are capable of producing spores, which are particular robust "dormant" cells that survive being incorporated into the concrete matrix for self-healing. These spores are not only adapted to be metabolically active in highly alkaline environments. One type of ureolytic bacteria that can adapt to the conditions of soil is *Bacillus megaterium*.

It has been discovered that the common soil bacterium *Bacillus subtilis* can cause silica precipitation and other metals to bind to its cell wall (Urrutia and Beveridge, 1994). These investigations demonstrate a model method for modifying cell walls with Bacillus species through the electropositive process of ethylenediamine to bind the SiO3-ion from the silicic acid. According to this method, after cement reacts with water during the primary reaction, it induces calcium silicate hydrated (C-S-H) gel and calcium hydroxide in aqueous form. The secondary hydration process will then proceed with the presence of pozzolans (i.e., siliceous and aluminous material) inside the concrete and produce an additional C-S-H gel that accelerates and enhances the final concrete's strength. The process of biomineralization, which forms precipitation and fills the pores that enhance the mechanical properties in concrete materials, such as strength and durability, can be naturally derived from microorganisms. In this case, using microorganisms as an alternative method could result in a precipitation that has characteristics similar to those of cement pozzolans. Thus, evidence of bacterial modification in the presence of silicate, whereby aqueous calcium hydroxide reacts to form C-S-H, and improvements made to the development of soil's compressive strength, warrant additional research in this area (Afifudin et al., 2011).

6.3.2. Bio-Enzymes and Their Role in Construction Engineering

Current experimental developments center on the use of protein engineering to modify soil biochemically. Since bio-enzymes can now be produced in larger quantities and are less expensive than traditional chemical stabilizers like bitumen, Portland cement, and hydrated lime, which are needed to stabilize soils, more research has been focused on using them as soil stabilizing agents. Seeing the "termite engineers" demonstrate their engineering prowess in building their all-weather termite mounds inspires creative thinking (Wijeyesekera, 2014). Meter-high, extremely hard soil structures are constructed using the enzyme-rich ant saliva. It is well known that these structures endure long, intense tropical rain seasons (Assam et al., 2016). As research in this area progresses, enzymes like peroxidases, phospholipase, cellulase, phytase, lipase, luciferase, and proteases are among the biochemical solutions for stabilizing soil. Enzyme-based soil stabilization shows that despite intense traffic and a lot of rain, enzymes have been effectively used to stabilize roads in Malaysia, China, India, and the Western United States (Taha et al., 2013). In addition to making the roads stronger and more durable, the project's cost has been decreased. The California Department of Transportation has tested an enzyme-based compaction additive in Mendocino County on multiple occasions. This natural product reduced dust and improved chip-seal applications by assisting the road base in setting very tightly. Dust reduction is mandated by agencies responsible for water and air quality, so this novel product has the potential to be effective and is less expensive than asphalt. Non-hazardous and environmentally friendly products like Permazyme and EcoBric use low-energy techniques for soil stabilization and have weather resistance that is comparable to that of concrete. Competitive bio-enzyme products, such as Renolith, Fujibeton, and Terrazyme, have the advantages of being quick, affordable, easy to use, and utilizing environmentally friendly technology to produce a "value added" product (Rajoria and Kaur, 2014). While these enzyme-based soil stabilizers seem to have many benefits over traditional chemical stabilizers, it is unclear how and when these products function. These products have limited applications for their stabilizing mechanisms, and it is challenging to forecast their long-term performance due to their proprietary chemical composition.

Ethoxylated alcohols, fermented vegetable extracts, and non-ionic surfactants make up the stated formulation of Terrazyme, according to Khan and Taha (2015). The manufacturer states that Permazyme is made up of several different enzymes, proprietary components, and an unidentified organic material. Additional laboratory analysis revealed that both preparations contained proteins and non-ionic surfactants (AbouKhadra et al., 2018). Soil roads can be rendered waterproof and hardened by applying a bio-enzyme treatment using various multi-enzyme formulations or in conjunction with other polymer compounds. It is crucial to conduct research that can provide unbiased scientific backing for the application of novel enzymes as a product to stabilize soils. Enzymes increase resistance to freeze-thaw cycling and provide a desired reduction in the compaction effort required to reach target soil densities in clayey soils (Rajoria and Kaur, 2014). The suggested soil types, however, are restricted to having a maximum liquid limit of 30%, a plastic index between 2 and 10, and a clay content of 12 to 30% with 18 to 30% cohesive fines.

The application of enzymes to products used to treat soil to improve horticultural applications gave rise to the concept of using enzyme stabilization for roads. A process change resulted in a material that can be used to stabilize unstable terrain for vehicular traffic. The enzymes improve the soil particles' ability to bond and wet when they are introduced. A cementing bond that stabilizes the soil structure is thought to have been formed by the enzyme or enzymes reacting with the molecules of the soil. Increased application rates have the potential to stabilize the treated soil and create a solid, compact, water-resistant bond layer suitable for use as a road surface.

A common description serves as the basis for preliminary laboratory research on different enzyme products as soil stabilizers: some enzymes are adsorbed by the clay lattice, which starts a catalytic reaction (i.e., exchange mechanism for metal cations). Cation exchange describes how the clay and enzyme react right away, within a few minutes of mixing, to improve the texture of the soil. They have a significant impact on the clay lattice, causing it to expand and subsequently contract. Additionally, colloids have the ability to absorb the enzymes, allowing them to pass through the soil electrolyte media. Enzymes help break down the lattice by assisting soil bacteria in releasing hydrogen ions, which causes pH gradients at the surfaces of clay particles. The soil becomes more workable and constructible due to the altered structure, which also reduces its sensitivity to moisture.

Furthermore, element stabilization and enzyme activity typically correlate. Metal ions, like calcium ions, are necessary for the upkeep of several enzymes' stable and functional structures. These ions limit local flexibility and unfolding because they are attached to particular binding sites on the molecules' surface. When present in excessive concentrations, elements with similar ionic properties can replace the trace amounts of specific divalent metal ions that enzymes require as functional cofactors. For example, Mn^{2+} is required for glucosyl hydrolase, Zn^{2+} is required for phosphohydrolase, and Ni^{2+} is required for amidohydrolase. In actuality, the catalytic contributions of enzymes and variable substrates—which provide microorganisms with energy—are primarily responsible for the biochemical reactions. By replacing the monovalent cations surrounding the surface of expansive clay soil particles, enzymes with Ca^{2+} binding domains or other metal binding sites, like Co^{2+}, Mg^{2+}, and Zn^{2+}, can act as a soil stabilizer through ion exchange and potentially reduce bound water molecules. As a result, the bonds and clay lattice structure are strengthened, and the clay particle flocculation is made possible.

Enzyme use is special because it's one of the main modern approaches to biochemical processes that save energy and materials while also being ecologically friendly. More novel enzymes must be found, and their mechanisms of reactivity with different soil geochemistry and/or wastes, like fly ash, must be clarified. The possibility of stabilizing waste material with enzymes raises the question of whether they can cooperate. Nonetheless, several enzymes have demonstrated promise in associated industrial uses, including protease, lipase, urease, cellulose, arylsulfatases, amylase, and chitinase.

Many stabilization mechanisms, known as "enzyme induced calcite precipitation" (EICP)—which is comparable to the process known as "microbial induced calcite precipitation" (MICP)—have been reported. These mechanisms include the activation of bacterial lipases by Ca^{2+} and other metal ions. Because of these characteristics, immobilized lipase is a desirable solution for applications involving soil stabilization (Carvalho et al., 2013). These Ca^{2+} activators might promote calcium adsorption on soil particles, which would free up metal ions to attach to clay soil particles. It is hypothesized that enzyme products with high silica content, such as silica-immobilized lipase or calcium and sodium, act more like cement when added to soil because they form hydrated calcium silicate.

Microbially induced phosphate deposition is the term used to describe bio-phosphate cements. In bacterial solutions, alkaline phosphatase continuously hydrolyzes the phosphate monoester, producing PO_4^{3-} ions. When PO_4^{3-}ions react with metal cations (alkaline earth element Ba), $MHPO_4$, MPO_4, or $M_3(PO_4)_2$ are created, which are then used to cement loose particles. Van der Waals bonds in sandstones are created when the cementing mechanism of barium hydrogen phosphate particles through microbial precipitation interacts with quartz sand to form sizable agglomerates. Loose sand particles can be consolidated into a bio-sandstone using bio-phosphate cement. Additionally, the effectiveness of soil enzymatic stabilizers for use in road applications draws attention to Terrazyme and Permazyme, two well-known commercial preparations. Overall, using bio-enzymatic soil stabilizers minimizes the need for hauling materials, lowers the amount of water needed at construction sites, and shows promise as an additive to lessen the application of OPC. When compared to traditional road construction methods, enzyme products are highly economical, saving three to five times as much due to factors like their non-toxic formulation and natural fermentation processes.

The kind of soil, the length of time it takes to cure, and the application technique all have a big impact on how well the bio-enzymatic treatment works. Peng et al. (2011) discovered that the UCS of the enzyme-treated soil was greater under air-dried curing conditions than OPC because the cement particles could not be fully hydrated in such conditions. The UCS of the bio-stabilized soils was less than 0.5 MPa under sealed curing conditions. Marasteanu et al. conducted triaxial tests to ascertain the resilient modulus of two types of soils—soil-1, which contained 96% of fines (75% clay) and soil-2, which contained 60% of fines (14% clay)—that had been treated by two distinct enzymatic preparations (1 cc of enzyme per 5 L of water). There were two confinement pressures used for the triaxial tests (27.6 kPa and 55.16 kPa). The findings demonstrated that adding enzyme-1 to the soil enhanced its shear strength by 9% and 23%, respectively, in soils 1 and 2. For soils 1 and 2, the increases in shear strength were 31% and 23%, respectively, in the case of enzyme-2 treatment. The treated soil's resilient modulus was between 86,453.4 kPa and 5,805.4 kPa, which was higher than the mean of the untreated soil. The tests were conducted using a deviatoric stress of 96.5 kPa and confinement pressure of 13.8 kPa. Soil stiffness is commonly defined by the soil's resilient modulus. For example, at 48.3 kPa deviator stress, soil-1 with enzyme-2 showed improved resilient modulus; the increase in resilient modulus ranged from 55 to 85%. For soil-2, the increase in resilient modulus at the same deviator stress level was 51% to 61% with enzyme-1 and 57% to 137% with enzyme-2.

For the fine-grained soils, enzymatic stabilization significantly improved UCS and CBR. The study's improvement of the enzyme product as a measure of soil strength was expressed by (CBR)r and normalized to that of untreated soil. The best concentration was 2.5 g/L. For fine-grained soil, the improvement in (CBR)r ranges from 2.75 to 4.5 times those values observed for non-treated soil. Additionally, as the amount of clay in the soil increased, the percentage of fines became more significant. Permazyme reached its maximum (CBR)r (4) at 0.5% concentration and 30–40% clay content. The UCS of the fines material was 800–1200 kN/m^2. The findings suggest that applying the enzymatic preparations greatly improved the fine-grained soil's strength, with the rate of improvement being correlated with the soil's clay content. Panchal and Associates (2017) also discovered that it was essential to use the proper dosages. For example, a 700 mL dosage of Terrazyme enzyme solution, which treated 1 m^3 of soil, consistently increased the soil's CBR values to 9.80% from an initial value of 8.31% (unsoaked) and 4.19% (soaked) after the first week of curing, 13.43% after the second week of curing, and 14.80% after the fourth week of curing. On the other hand, CBR values significantly varied with longer curing times and higher dosages (900–1,000 mL/m^3). Terrazyme, a class of bio-catalysts consisting of water, ethoxylated alcohols, and matured vegetable concentrate, has demonstrated its effectiveness in modifying the strength, consistency, and other geotechnical properties of fine-grained soils. To get the most out of this bio-enzyme additive, soil gradation must be met with the minimum amount of clay needed for the reaction, enough active elements to react with metal cations and break down bonds sequentially, and enough advancement of a new link system.

The features of the soil mixtures also determine how differently the enzyme is broken down. Because the strength improvement decreases with longer curing times, a crucial component of the biocatalyst is inadvertently overlooked in that study. Because the enzymes do not weaken during the regeneration of the enzyme catalyst after the reactions are finished, as suggested by Thomas and Rangaswamy in 2019, low concentrations of enzymes are sufficient for the improvement of soil. The authors presented an optimal soil mixture consisting of clay soil supplemented with 1% cement and 0.06–0.06 mL/kg of Terrazyme, which showed increased UCS. The strength of clay soil was 28 kPa at first, but it increased to 127 kPa with the addition of 1.0% cement and 0.06–0.6 mL/kg of bioenzyme. This is approximately 354% more than the UCS of untreated soil. Longer curing times could maintain this increased strength. Shifts in the FT-IR results and the formation of extra CS-H gel from the SEM images corroborated these changes in the strength properties of the soil matrix.

Due to their inconsistent performance and frequent name changes, the majority of enzyme stabilizers on the market are imported from outside of South Africa. As a result, the local road construction industry has not given these stabilizers much support. The construction industry is investigating the use of biological methods to improve structural properties like strength, volume stability, durability, and permeability in road construction applications because there aren't many creative solutions using conventional approaches. The inability to perform has led to the identification of a gap in the production of better bio-stabilizers for the regional market.

6.3.3. Role of Biopolymers in Construction Engineering

Fossil fuels have supplied the resources for the polymer industry for more than 50 years. But it's a fact that petroleum resources will eventually run out. Due to the substantial depletion of fossil fuels, sustainability has gained importance, and influential figures in science and politics have pushed for the paradigm shift to "bio-based material." The 1920s saw the introduction of lignosulfonate, a biopolymer, for OPC concrete plasticization, marking the beginning of the twentieth century's era of admixtures. This was the first functional polymer to be widely used in the building industry. Although there has been a significant increase in publications on the use of bio-based polymers in recent years (Chang and Cho 2019), the polymer industry still only uses a small portion of these materials. According to Averous and Pollet (2012), biopolymers are polymers derived from agro-resources (polysaccharides, cellulose, starch, chitin, chitosan, and alginates); microorganisms through fermentation (polyhydroxyalkanoates, like polyhydroxybutyrate); and biotechnology through conventional synthesis (polylactides (PLA), polybutadiene succinate, biopolyethylene (PE), polytrimethylene terephalate, and poly-p-phenylene). Reusing biomass and agricultural waste will increase the environmental benefits of biopolymers over conventional petroleum-based polymers.

Naturally occurring polymers with polymeric carbohydrate structures are called biopolymers. The kinds that are currently utilized in the building sector are gellan gum, xanthan, guar gum, welan, curdlan, chitosan, and sodium gluconate. To increase viscosity, water retention, set retardation, and flow ability, these are utilized in injection grouts, wall plasters, self-leveling under layers, and dry-mix mortars (Plank 2004). It has been demonstrated that the biopolymer b-1,3/1,6-glucan aggregates soil particles, increasing the soil's overall compressive strength. Additionally, it has a negative impact on the consolidation coefficient and a positive impact on the treated soil's Atterberg limits, swelling index, and compactability. By improving soil inter-particle cohesion, a small amount of biopolymer significantly reduces soil erosion

(Fatehi et al., 2018). Research on guar gum and xanthan by Ayeldeen et al. (2017) demonstrates exceptional soil fortification for use in the road sector. Conversely, Chang et al. (2015) employed two thermos-gelation biopolymers (gellan gum and gaur gum) to improve the strength and durability of sandy and clayey soils by enabling hydrogen bonding. Temperature was a significant factor in soil treatment, but time was not as important. Further research is necessary since the degree of soil improvement varies depending on the types of polymers used, the types and composition of the soil, the doses used, and the curing conditions.

These studies offer a sustainable way to enhance the geotechnical behavior (strength) of various soil types, which may eliminate the need to specify intricate guidelines for microbial injection techniques. It is possible to compare the financial viability of soil-biopolymer matrix interactions and associated soil-strengthening mechanisms to regular Portland cement (OPC) in the field. Chang and Cho (2019) state that when compared to conventional soil treatment (OPC), the cost of treating soil with biopolymers like gellan gum is still somewhat high. They demonstrated that, in contrast to the current cost of USD 10 for a 10% cement treatment, the total cost of treating one ton of soil with gellan gum via thermal treatment is about $25–30 USD. However, because of the rapid growth of the global biopolymer market, it is becoming more feasible to use biopolymers for construction purposes, according to recent analyses of economic feasibility and prospects. Results also show that, in the context of environmentally friendly building and development, biopolymers with high tensile strength, such as bioplastic PLA, have a strong potential to replace cement as a soil treatment material.

The use of affordable stabilizers on readily available soils (in situ) that are both economical and ecologically benign is becoming increasingly necessary. The development of bio-based stabilizers has a lot of promise, but more study is needed to test the idea and create product prototypes that can be used with *in-situ* building materials.

6.3.4. Micro-Level Analysis to Understand the Mechanism of Interaction of Soil Stabilizers

Subgrade soil is enhanced in its engineering properties by non-conventional additives. The macro-level strength characteristics of these products are determined by an evaluation that is based on a combination of *in-situ* standard performance testing. Studies, however, indicate that it is not always possible to forecast the anticipated macro-scale performance. As a result, additional complementary techniques must be found for the design. Therefore, determining a mechanism of action through the use of micro-structural details may help to identify how the product interacts with the soil particles.

For geotechnical engineering applications, the physico-chemical bonding mechanisms of stabilized material induced by stabilization processes are advantageous. Through micro-level analysis, mechanistic strategies such as bioclogging, bioaggregation, and bio-cementation can be identified. Latifi (2016) employed X-ray powder diffraction (XRD), field emission scanning electron microscopy (FESEM), energy-dispersive X-ray spectroscopy (EDX), Fourier transform infrared spectroscopy (FTIR), Brunauer, Emmett, and Teller (BET) surface area analysis, and particle size analysis (PSA) using a laser diffraction approach to demonstrate the mechanisms of pore filling and particle bonding by cementitious products. Together with these tests, unconfined compressive strength (UCS) tests demonstrate that calcium-based additives, such as the commercial product *SH-85* (Latifi et al., 2013), improve strength. To evaluate prototype bioadditives for *in-situ* soil stabilization, morphological representations can be determined using micro-level analysis techniques like *FESEM* (Mgangira and Ndibewu, 2010). Given that the modes of action of biological stabilizers are not mutually exclusive and that bioadditives may have one or multiple functional effects, it is crucial to determine the degree of laboratory performance testing required, in addition to large-scale testing, to analyze the interaction between soil particles.

6.4. ADVANTAGES OF BACILLUS SPECIES IN INDUSTRIAL APPLICATIONS

The quality of the soil subgrade, or more specifically, whether or not the subgrade has a low bearing capacity, which can cause premature distress and failure of the newly constructed road, is a problem that arises frequently in road construction. A recent study by Samang et al. (2018) demonstrated the use of *B. subtilis* to improve the bearing capacity of subgrade soil, indicating that bio-stabilization with microbes is one way to address this issue.

The theory suggests that Bacillus species (i.e., *Bacillus subtilis* and *Bacillus cereus*) generate hydrolase enzymes, such as lipases, which function in concert with additional ionic stabilizers to potentially enhance the quantity and strength of ionic bonds on the soil-clay particle surface. As a result, a less expansive and more highly weathered clay structure is formed. Bacteria produce four main classes of polymers: inorganic polyanhydrides (like polyphosphates), polysaccharides, polyesters, and polyamides. It is suggested that the enzymatic stabilizers and biopolymers will create polymer-soil matrices because of the ionic and hydrogen bonds that bind the soil-clay particles together. As they combine with the natural polymers to form both inorganic and organic catalysts, the enzymes in this instance will play a dual role in polymer stabilization, strengthening the structural properties of the soil. As a result, the soil material is much less moisture-sensitive, more workable, and can be compacted by machinery and traffic forces to a better particle-interlock state. Increased internal friction and enhanced bearing capacity are the results of better particle-interlock (Firoozi et al., 2017).

Bacillus species: their significance: *B. subtilis* in the synthesis of surfactants, including fatty acids, polysaccharide-protein complexes, glycolipids, phospholipids, and lipopeptides.

These substances allow for increased reduction of surface and interfacial tension, a decrease in the critical micelle concentration, improved emulsification qualities, and resistance to ionic strength. The stabilizing effect of extracellular polymeric substances (EPS) made by both *B. licheniformis* and *B. subtilisis* also suggested by recent developments in soil research. Toxic metals can be bound by EPSs, which can also facilitate mineral dissolution and improve the aggregation of biofilms. They can interact with other solutes and soil particle surfaces when released into a solution, which could help stabilize the microbially produced organic carbon in the soil. In general, one can anticipate that these microbial products will improve the weak or unstable subsoil particle's ability to bind.

Bacillus species constituents. Ferment is a well-known term for the extracellular polysaccharides produced by bacteria when they have access to too many carbohydrates or other water-soluble carbon sources relative to nitrogen sources. Bacterial water-insoluble polysaccharides are produced industrially using food processing wastes or byproducts like corn syrup, cassava syrup, and molasses with a carbon-to-nitrogen ratio greater than 20. Biofilms, which are communities of bacterial cells living in a sticky extracellular matrix that they produce on a liquid or solid surface, are formed in large part by extracellular polysaccharides. Similar to many other bacterial phenotypes, the ability to form biofilms is the consequence of transcriptionally distinct cell types in bacteria differentiating. In this sense, the capacity for cellular differentiation in the model organism *B. subtilis* has been thoroughly studied.

A mixture of polysaccharides, amino sugars, proteins, teichoic, and nucleic acids makes up the slime that is released into the extracellular medium by the extracellular particles (EPSs). The EPSs can also remain attached to the cell surface. EPS can be heteropolysaccharides or homopolysaccharides, and it can be embellished with phosphates, sulfates, acetyl groups, and N-acetyl-amino sugars, among other residues. The literature reports the presence of flocculation activity in bacteria that produce different types of polysaccharides, specifically *B. licheniformis* and *B. polymyxa*. The EPS serves as the biological active biofilm matrix's "construction material" and has multiple uses, including gel formation, flocculation, emulsification, absorption, film formation, and protection.

The Bacillus species are the most appropriate candidates that exhibit promise in generating extracellular polysaccharides to bind soil particles and fill in soil pores. Studies reveal that bacteria that produce polymers grow in soil, and the geotechnical uses of these bacteria may act as a long-term stabilizer. Thus, it is broadly relevant to comprehend the chemical signals that cause bacteria to form biofilms and produce extracellular polymers. Numerous Bacillus species within two days of cultivation, displayed the highest flocculation activity. The flocculation was caused by the release of the polysaccharide from the cell surface into the culture medium, which binds the kaolin clay particles. After being released into the soil, EPS may aggregate with polyvalent metals or sorb to mineral surfaces, which can enrich microbially produced proteins and carbohydrates in the clay fraction or in high-density mineral organic associations. Furthermore, EPS has the potential to both neutralize the surface charge of the bacteria and make the surfaces of the cells more hydrophobic. The saturated hydraulic conductivity of sand can only be significantly reduced by slime producers, according to a study conducted on four bacterial strains that were used to inoculate quartz sand packed in columns with a continuous supply of a medium. One potential source of inexpensive microbial polymers is the sewage sludge produced annually by municipal wastewater treatment plants, which is a waste product of microbial biomass. Sewage sludge production amounts to several million tons.

The utilization of biological methods in geotechnical engineering necessitates a fundamental comprehension of the mechanisms underlying frequently problematic materials, including their identification and measurement. This is required to put protective measures in place against the harm that incorporating such materials causes to road structures. These issues, such as soft soil subgrade with a low bearing capacity, are frequently encountered during the building of roads. It is hypothesized that bio-stabilization techniques, or the use of bacteria, can solve issues like soft clay. Research conducted with B subtills slurry revealed that the highest CBR value measured on soft soil was 39% after a 7-day curing period and a 6% bacterial solution. This value was 13 times higher than that of the untreated soil. Nevertheless, the CBR value decreased as the bacterial treatment concentration increased (8–10%). This initial assessment demonstrated that subject to obtaining the proper dosage, the application of *B. subtilis* bacterial solution on soft soil greatly increased the soil's bearing capacity as a potential subgrade layer. Improvement in bearing capacity was also demonstrated with high tensile polymers generated by these microbiological constituents. A dual effect from *B. subtilis* based on (b) biofilm formation in the presence of soil clay minerals, induces attraction forces between microorganisms and soil particles that later develop into larger irreversible binding properties, was identified by this method of biological treatment. The two-pronged effect includes (a) microbial injection and by-product precipitation and may be introduced as a new type of soil binder.

6.5. ASSESSING THE PERFORMANCE EFFECT (MODES OF ACTION) OF ADDITIVES FOR STRUCTURAL APPLICATIONS

Expansive clay, a type of soil material that is widely distributed throughout the world, can cause significant harm to road infrastructure because of variations in its volume caused by variations in water content. Various chemical products, also known as stabilizers or additives, are employed to increase the strength of soil. The process of stabilizing these soils involves the use of conventional additives like fly ash, cement, and lime. The flocculation of the clay layers is caused by additives containing calcium oxide, which causes the monovalent ions to be replaced with Ca^{2+} ions. This lessens the electrochemical charges and electrochemical forces of repulsion between the layers of clay minerals and balances the electrostatic charge of the clay's layers. A soil with improved engineering properties (a more granular structure, lower plasticity, greater permeability, and above all lower expansibility) is created when clay particles adhere to one another to form flocs.

The outcome is still unsustainable from an environmental standpoint, even if weak subsoil material replacement can be minimized. Thus, bio-based methods for reasonably and appropriately modernizing roads have recently been adopted in numerous nations. These methods maximize the use of locally available, in-situ resources. The direct structural performance/effect (modes of action) of commercial products containing enzymes, polymers, or other proprietary biological road additives, however, is the subject of a dearth of literature. This is especially problematic for the road industry because these products can be used selectively depending on the type of soil and have complex application rates or mechanisms. In the past, road and civil engineers in South Africa have shown reluctance to effectively standardize or implement these liquid chemicals in road sector applications.

Although difficult to predict, a substantial amount of research and development (R&D) has been conducted to propose mechanisms of non-traditional additives. However, the majority of these studies have several limitations, including those related to application rates, concentrations, soil properties, and plasticity index. As a result, it has been challenging to standardize functional aspects regarding the mechanism of non-traditional additives. They have proven a national standard (*ex-situ*) method for figuring out the strength properties of additives based on polymers and enzymes. This approach infers the modes of action and associated mechanisms may be attributed or demonstrated by a pre-defined strength criterion and to the performance effect of key bio-molecules (i.e., resistance to compression, erosion, abrasion, and water absorption), as described below:

6.5.1. Bio-Stabilizer Effect: Resistance to Compression Load

A material's ability to withstand applied stress is a sign of its quality and structural integrity. Examining the compression load implies evaluating the material's ability to withstand deformation, making it a crucial evaluation of material strength to guarantee stability and long-term durability. For instance, the compressive strength of bio-cemented sand (calcite) increases in a uniaxial compression test. The increased durability of this material is adequate to meet geotechnical applications. Furthermore, the addition of auxiliary additives like fly ash, lignosulfonate, and biopolymers has been shown to increase compression strength. Compression testing is a crucial design criterion for a quantifiable assessment of the impact of a bio-based stabilizer or additive on performance, according to supported laboratory and field application research.

6.5.2. Bio-Stabilizer Effect: Resistance to Erosion

When the shear pressure exerted by flowing fluids surpasses a certain threshold, surface erosion takes place. Material failures of geo-structures are frequently caused by this surface erosion. Increased soil erosion resistance has been achieved through a variety of strategies, including chemical ones. Ham et al. (2018) demonstrate how the use of microbial biopolymers can improve the resistance to surface erosion while modifying the mechanical and hydrological properties of soils, including inter-particle bonding, stiffness, strength, and permeability due to the insoluble biomaterials. Biomaterials have the potential to improve erosion resistance through various mechanisms. Firstly, gel-like EPS (polymers with hydrophobic properties) can fill soil pores and reduce hydraulic conductivity, causing bio-clogging. Secondly, film-like biopolymers can increase apparent cohesion and erosion resistance through grain coating and bridging. Lastly, the bio-cement method can also be used to control soil erosion caused by hydraulic scour which increases resistance to erosion by using cement, xanthan, gellan gum, and lignosulfonate. When added to sands, biopolymers and clays have a similar effect on erosion control behavior by making the soils appear more cohesive and lowering permeability and void ratio. The material's ability to withstand flowing water and simulate rainy conditions is tested in many ways, including the erosion test, which is a preliminary test required for additional testing.

6.5.3. Bio-Stabilizer Effect: Resistance to Abrasion

When it comes to unpaved or earth roads, the most common type of deterioration depends on whether the weather is wet or dry. In addition to being resistant to excessive dust during dry weather and excessive slipperiness during wet weather, a good wearing course on these roads should be able to withstand the abrasive action of traffic and erosion by water and wind. All unpaved and earth roads experience surface abrasion loss during dry conditions, which reduces soil particle adhesion and generates dust. The movement of loose material causes corrugations under traffic action and raveling of the road surface when there is not enough binding power of the material to keep the surface intact. The excessive amounts of dust produced by these unpaved roads impair traffic safety and visibility. In most developing nations, over 90% of the road network consists of unpaved routes, which account for a sizable portion of the country's total road network (Burrow et al., 2016). For this reason, most people have access to unpaved roads, which also promote social and economic development. Unpaved roads create many issues, such as low standards, rough riding, increased erosion during rainy seasons, and copious amounts of dust from passing traffic and wind. According to Molozane and Visser (2014), one of the contributing factors is that many unpaved roads lack the proper range and distribution of particle sizes, which would provide a good load-bearing capacity when wet (coarser particles) or enough plastic capability (clay) when dry to prevent material from breaking loose.

According to historical data, in 2004 there were about 3 million tons of dust produced annually on unpaved/unsealed roads in South Africa. According to estimates, two-thirds of this dust settles on the road and the million tons that remain are lost forever. This causes a detrimental effect on the environment and accelerates deterioration because of material loss (Jones and Ventura, 2004). The occurrence of potholes and ruts on unpaved roads leads to a reduction in their strength and durability, necessitating replacement and regular maintenance. The depletion of natural resources makes it unsustainable to replace gravel material. Applying soil stabilizers (additives) could be one way to address this growing issue. As was previously mentioned, conventional additives like cement and lime can be used to improve road materials. However, research on ionic additives, such as enzymes and sulfonated products, offers a workable and affordable solution to persistent issues (Moloisane and Visser, 2014). It is crucial to look into dust suppressors, also known as dust palliatives (such as Soil Seal and Aggrebind products), which are substances applied to the surface of the road to stop the development of dust clouds and the effects they have on the environment.

6.5.4. Bio-stabilizer Effect: Resistance to Water Absorption

Because unpaved roads are made of soil that is more porous and permeable and has the potential to expand and lose cohesiveness, they are more susceptible to weather events, particularly during rainy seasons. In arid conditions, chemically or biologically stabilized soils might have sufficient compressive strength, but when the moisture content is unfavorable, they become weaker. Therefore, in this instance, the amount of water absorbed by the stabilized soil is significant. The presence of additives (polymers, which have hydrophilic qualities) in soil is a major factor in lowering soil permeability because they increase the viscosity of pore water and decrease pore volume. The biofilm method can be used to address seepage issues like the decreased leakage volume of earth dikes and dams. Field experiments have been carried out using a bioclogging process. In a different field experiment, wind forces and heavy metal-contaminated soil water seepage were remedied using the bio-cement grouting method (Cuthbert et al., 2013). Water-repellent chemicals are therefore thought to be a useful way to stop moisture damage in building materials.

Hodges and Lingwall (2020) have examined the bio-augmentation – enzyme induced precipitation technology in the field. They employed three sizable field test sites with an emphasis on dust suppression and road surface erosion mitigation. A test site featuring weathered shale demonstrated the bio-stimulators' potential for mitigating and preventing surface erosion. This resulted from the clay content of the soil characteristics as opposed to the sandy soils. The primary finding of their study was that it had multiple implementation flaws in the field. For example, microbial bio-mineralization may not happen at all in soil with high salt concentrations (i.e., high salt concentrations kill bacteria if the bacteria's tolerance to chlorine is exceeded). A study of this case study led to the suggestion of two research questions for the development and performance of future bio-stabilizer products: (i) Will a different formulation (using bacterial carriers) be ideal for installing microbial-based bio-stabilizer treatment on-site, as opposed to using live cultured bacterial suspensions on-site or transported from laboratories?; and (ii) Will longer curing times, like 28 days, be optimal as this is directly dependent on strength gains and positively affects sorption, coating, and the aggregation effect of the ingredients by the stabilizer?

This literature review sought to address issues related to road construction challenges. Through the use of microbial components and produced metabolites, CSIR (SA) has identified many mechanisms of action for enhancing the strength properties of construction materials (i.e., bio-enzymatic, surfactants, ionic, and polymer soil stabilization). To achieve a specific performance requirement, such as workability, mixability, binding, and compaction, this enables the proper isolation and assessment of strains based on their capacity to produce unique compounds of interest. This leads to longer-lasting road surfacing, improved road substructure, and a reduction in the need for ongoing maintenance. The study is significant because it describes critical steps in the creation of products made from Bacillus species that stabilize soil and associated bio-additives for use in road construction. These bio-stabilizer products, which aim to address challenges in road construction, are representative of fermentation components, whole cell bacterial suspension, and biopolymers. Currently, research is concentrated on creating new products from Bacillus species. There is not much fermentation. Therefore, the goal of this literature review was to gather information about and assess the potential application of fermentation products, such as biopolymers, active whole cells, and "sticking" agents made by microbes, in the processes of stabilizing soil. A summary of key focus areas for development of bio-stabilizer product research:

- To evaluate the essential structural properties of biological components or fractions and subsequently identify the best performers, medium- to high-throughput screening is being developed.
- Large-scale demonstration of functional components and verified selection of candidate fractions.
- The candidate prototypes' mode of action, which involves structural criteria like resistance to abrasion, erosion, water absorption, and compression load, should be demonstrated.
- Top prototypes are put to the test in actual road subsurface investigations. Performance will be demonstrated through laboratory-based simulated tests; using the Heavy Vehicle Simulator is also advised.

6.6. OTHER ENVISIONED APPLICATIONS: ENVIRONMENTAL

There is a strong global push to support low-carbon, environmentally sustainable economies. Finding strategies that support this shift is necessary. Repurposing solid waste to create soil-based construction materials is one of the many economically and technically viable approaches to waste management. The amount of waste generated daily in South Africa is 54,425 tons, which ranks 15th globally. The fact that South Africa's landfill areas are rapidly filling up is not surprising. Investigating different uses for bio-based products is part of the process of developing a novel bio-based technology. One such application is the possibility of enhancing the mechanical characteristics of feeble soils through the addition of industrial waste and byproducts like fly ash.

Many waste materials can be converted into useful products, and this is a suggested first step since these methods maximize the use of locally available resources. By concentrating on improving the unique mechanical and physical characteristics of soil, high-energy waste materials like ash, plastic, paper, or agricultural waste (such as sugarcane bagasse) can be used as a binding agent to address the many difficulties encountered in soil stabilization. e. enhanced durability against repeated wetting and drying, hydraulic conductivity, and resistance to erosion or abrasion, as well as for environmental revitalization). Understanding the demands of the construction sector and the mutually beneficial relationship between product development and research are critical to establishing these new waste utilization opportunities. Finding low-cost, high-volume applications and turning waste into products with value additions are the challenges.

Many waste materials can be converted into useful products, and this is a suggested first step since these methods maximize the use of locally available resources. By focusing on improving a soil's unique physical and mechanical properties—such as its resistance to erosion or abrasion, hydraulic conductivity, durability against repeated wetting and drying, and potential for environmental revitalization—the use of high-energy waste materials—such as ash, plastic, paper, or agricultural waste—as a binding agent can help address the many challenges faced in soil stabilization. Understanding the demands of the construction sector and the mutually beneficial relationship between product development and research are critical to establishing these new waste utilization opportunities. Finding low-cost, high-volume applications and turning waste into products with value additions are the challenges.

6.6.1. Industrial Wastes as Proposed Auxiliary Additives for In-Situ Soil Stabilization

6.6.1.1. Ash Waste

The boilers' bottom ash is collected, i.e., fly ash, an inflammable, coarse, granular byproduct of burning coal, is extracted from the gas washers (i.e., the majority of the unburned material from the pulverized coal that burns in boiler bottoms is captured in flue gas. Ash is currently a big issue; according to Gomes et al. (2020), fly ash production is about 46 million tons annually worldwide. Fly ash has a higher proportion of organic matter than bottom ash, which is made up of both organic and inorganic components. Fly ash contains heavy metals that are not biodegradable; instead, they can only change chemical states, which alters their toxicity and mobility. Ash has the potential to be used in more advanced ways, such as deep soil application, but there are still questions about its toxicity and environmental leaching (Kumpiene et al., 2008). One suggestion for combating this is the application of microorganisms. Microorganisms can affect metals in several ways:

- they can transform some forms of metals through a redox reaction or alkylation;
- they can accumulate metals through metabolism-dependent (passive) or metabolism-independent (passive) uptake within cells;
- they can indirectly affect metal mobility by affecting pH; and
- they can produce or release compounds that alter the mobility of the metals.

The suggested biomolecules, or EPS, came from different species of Bacillus and/or in concert with other green composites are suggested as possible ash waste carriers and stabilizing agents.

6.6.1.2. Sugarcane Bagasse

Produced in vast quantities as a by-product of the sugar manufacturing industry, sugar cane bagasse has economic significance. South African sugar mills produce 24.7 million tons of sugarcane annually, and every 10 tons of crushed sugarcane yields 3 tons of bagasse (Muniyasamy et al., 2016). The fibrous residue left over after sugarcane juice is extracted is called bagasse. About 50% of sugarcane bagasse is made up of cellulose, 25% is hemicellulose, and 25% is lignin. On the other hand, silica-rich sugarcane bagasse ash can function as an efficient pozzolan, enhancing pozzolanic reactions and producing better-performing building materials (James and Pandian, 2017). Many uses for these solid waste materials, including adsorbents, filters, ceramics, briquettes, bricks, and soil modifiers, have demonstrated their economic significance. Furthermore, because of their high tensile strength, permeability qualities, and ability to remediate soil, bio-plastics, or PE, a relatively new material made from bagasse and maize stalks, may play a part in road construction in the future (Muniyasamy et al., 2016). Regarding the biodegradable aspect of bio-based materials, tests are difficult and dependent on the environment in which they are conducted. Nevertheless, standards, such as those established by the International Organization for Standardization (ISO) and the American Society for Testing and Materials (ASTM), can be used to adapt and test green composite materials in soil environments to manage their end-of-life cycle.

For building purposes, it is noteworthy that not all biodegradable plastics are recyclable instead of biodegradable.

6.6.1.3. Lignin-Based Waste

A byproduct of the paper and lumber industries is lignin. Similarly, lignin can withstand biodegradation in an anaerobic environment and be left in a landfill for a very long time. Wood is made up of cellulose, hemicellulose, lignin, and extractives. After cellulose, lignin is the most prevalent organic polymer on the planet. Reusing lignin as a stabilizing additive for soil or in civil engineering applications like road embankments and dam foundations are sustainable uses for this organic by-product derived from biosynthesis. Another industrial by-product being researched as a soil stabilizer is lignosulfonate (LS). A polymer with a lignin-base that contained hydrophilic groups—alcohol hydroxyl sulfonate and phenyl hydroxyl—and hydrophobic groups—carbon chains—was able to improve the properties of soil. The stabilizing effect of the LS-treatment results in a decrease in the soil's Plasticity Index (PI), increased stiffness, and a UCS that is not brittle. The increase in strength properties is ascribed to the electrostatic reaction between the LS-water mixture and soil particles, which results in the PI changes. These changes may be caused by changes in the lattices of the soil minerals, such as a decrease in the crystalline size of the double layer thickness of clay minerals, and improved flocculation properties, which enhance the stabilized soil strength and plasticity. Because LS is a byproduct of other processes, it can compete with other chemical stabilizers and is reasonably inexpensive, having been produced yearly in large quantities (50 million tons) worldwide.

REFERENCES

1. AbouKhadra, A., Zidan, A. F., & Gaber, Y. (2018). Experimental evaluation of strength characteristics of different Egyptian soils using enzymatic stabilizers. *Cogent Engineering, 5*, 1517577.
2. Achal, V., Mukherjee, A., & Reddy, M. S. (2011). Effect of calcifying bacteria on permeation properties of concrete structures. *Journal of Industrial Microbiology & Biotechnology, 38*, 1229–1234.
3. Afifudin, H., Nadzarah, W., Hamidah, M., & Hana, H. N. (2011). Microbial participation in the formation of calcium silicate hydrated (CSH) from *Bacillus subtilis*. *Procedia Engineering, 20*, 159–165.
4. Alamgir, M., Campbell, M. J., Sloan, S., Goosem, M., Clements, G. R., Mahmoud, M. I., & Laurance, W. F. (2017). Economic, socio-political, and environmental risks of road development in the tropics. *Current Biology, 27*, R1130–R1140.
5. Alazigha, D. P., Indraratna, B., Vinod, J. S., & Heitor, A. (2018). Mechanisms of stabilization of expansive soil with lignosulfonate admixture. *Transportation Geotechnics, 14*, 81–92.
6. Alzard, M. H., Maraqa, M. A., Chowdhury, R., Khan, Q., Albuquerque, F. D., Mauga, T. I., & Aljunadi, K. N. (2019). Estimation of greenhouse gas emissions produced by road projects in Abu Dhabi, United Arab Emirates. *Sustainability, 11*, 2367.
7. Babu, R. P., O'Connor, K., & Ramakrishna, K. (2013). Current progress on bio-based polymers and their future trends. *Progress in Biomaterials, 2.*
8. Bang, S. C., Bang, S. S., Choi, S. R., Lee, S. J., Lee, J. H., & Kim, J. S. (2015). Method of manufacturing soil reinforced by microbe-based bio-binders and soil produced by the method. *U.S. Patent No. 9,074,134*.
9. Blauw, M., Lambert, J., & Latil, M. N. (2009). Biosealing: A method for *in situ* sealing of leakages. In *Proceedings of the International Symposium on Ground Improvement Technologies and Case Histories* (pp. 125–130).
10. Boutique, J. P., Baeck, A., Vanwyngaerden, N., Stonehouse, J., Gizaw, Y., & Rothgeb, T. (2007). Enzyme stabilizer: Google Patents.
11. Bryson, S., & El Naggar, H. (2013). Evaluation of the efficiency of different ground improvement techniques. In *Proceedings of the 18th International Conference on Soil Mechanics and Geotechnical Engineering.*
12. Burrow, M. P. N., Evdorides, H., Ghataora, G. S., Petts, R., & Snaith, M. S. (2016). The evidence for rural road technology in low-income countries. *Proceedings of the Institution of Civil Engineers–Transport, 169*, 366–377.

13. Carvalho, N. B., Barbosa, J. M. P., Oliveira, M. V. S., Fricks, A. T., Lima, Á. S., & Soares, C. M. F. (2013). Biochemical properties of *Bacillus sp.* ITP-001 lipase immobilized with a sol-gel process. *Química Nova, 36*, 52–58.
14. Chang, I., & Cho, G. C. (2019). Shear strength behavior and parameters of microbial gellan gum-treated soils: From sand to clay. *Acta Geotechnica, 14*, 361–375.
15. Chang, I., Im, J., & Cho, G. C. (2016). Introduction of microbial biopolymers in soil treatment for future environmentally-friendly and sustainable geotechnical engineering. *Sustainability, 8*, 251.
16. Cuthbert, M. O., McMillan, L. A., Handley-Sidhu, S., Riley, M. S., Tobler, D. J., & Phoenix, V. R. (2013). A field and modeling study of fractured rock permeability reduction using microbially induced calcite precipitation. *Environmental Science & Technology, 47*, 13637–13643.
17. DeJong, J. T., Fritzges, M. B., & Nusslein, K. (2006). Microbially induced cementation to control sand response to undrained shear. *Journal of Geotechnical and Geoenvironmental Engineering, 132*, 1381–1392.
18. Dejong, J. T., Mortensen, B. M., Martinez, B. C., & Nelson, D. C. (2010). Bio-mediated soil improvement. *Ecological Engineering, 36*, 197–210.
19. Dhami, N. K., Reddy, M. S., & Mukherjee, A. (2012). Improvement in strength properties of ash bricks by bacterial calcite. *Ecological Engineering, 39*, 31–35.
20. Fatehi, H., Abtahi, S. M., Hashemolhosseini, H., & Hejazi, S. M. (2018). A novel study on using protein-based biopolymers in soil strengthening. *Construction and Building Materials, 167*, 813–821.
21. Firoozi, A. A., Olgun, C. G., Firoozi, A. A., & Baghini, M. S. (2017). Fundamentals of soil stabilization. *International Journal of Geo-Engineering, 8*, 26.
22. Gao, Y., Hang, L., He, J., & Chu, J. (2019). Mechanical behaviour of biocemented sands at various treatment levels and relative densities. *Acta Geotechnica, 14*, 697–707.
23. Ghosh, P., Mandal, S., Chattopadhyay, B., & Pal, S. (2005). Use of microorganism to improve the strength of cement mortar. *Cement and Concrete Research, 35*, 1980–1983.
24. Gomes, H. I., Funari, V., & Ferrari, R. (2020). Bioleaching for resource recovery from low-grade wastes like fly and bottom ashes from municipal incinerators: A SWOT analysis. *Science of The Total Environment, 715*, 136945.

25. Guncheva, M., & Zhiryakova, D. (2011). Catalytic properties and potential application of *Bacillus* lipases. *Journal of Molecular Catalysis B: Enzymatic, 68*, 1–21.

26. Ham, S. M., Chang, I., Noh, D. H., Kwon, T. H., & Muhunthan, B. (2018). Improvement of surface erosion resistance of sand by microbial biopolymer formation. *Journal of Geotechnical and Geoenvironmental Engineering, 144*, 06018004.

27. Hodges, T. M., & Lingwall, B. N. (2020). Laboratory study in the treatment of burned soils with microbial augmentation for erosion control. In *Geo-Congress 2020: Biogeotechnics* (pp. 20–28). Reston, VA: American Society of Civil Engineers.

28. Hottle, T. A., Bilec, M. M., & Landis, A. E. (2013). Sustainability assessments of bio-based polymers. *Polymer Degradation and Stability, 98*, 1898–1907.

29. Igiri, B. E., Okoduwa, S. I., Idoko, G. O., Akabuogu, E. P., Adeyi, A. O., & Ejiogu, I. K. (2018). Toxicity and bioremediation of heavy metals contaminated ecosystem from tannery wastewater: A review. *Journal of Toxicology, 2018.*

30. James, J., & Pandian, P. K. (2017). A short review on the valorisation of sugarcane bagasse ash in the manufacture of stabilized/sintered earth blocks and tiles. *Advances in Materials Science and Engineering, 2017.*

31. Katz, L., Rauch, A., Liljestrand, H., Harmon, J., Shaw, K., & Albers, H. (2001). Mechanisms of soil stabilization with liquid ionic stabilizer. *Transportation Research Record, 50–57.*

32. Khan, T. A., & Taha, M. R. (2015). Effect of three bioenzymes on compaction, consistency limits, and strength characteristics of a sedimentary residual soil. *Advances in Materials Science and Engineering, 2015.*

33. Larson, S., Newman, J. K., O'Connor, G., Griggs, C., Martin, A., Nijak, G., Lord, E., Duggar, R., & Leeson, A. (2013). Biopolymer as an alternative to petroleum-based polymers to control soil erosion: Iowa Army Ammunition Plant. *Environmental Security Technology Certification Program Office (DOD) Arlington, VA.*

34. Latifi, N., Meehan, C. L., Majid, M. Z. A., & Horpibulsuk, S. (2016). Strengthening montmorillonitic and kaolinitic clays using a calcium-based non-traditional additive: A micro-level study. *Applied Clay Science, 132*, 182–193.

35. Ma, W., Peng, D., Walker, S. L., Cao, B., Gao, C. H., Huang, Q., & Cai, P. (2017). *Bacillus subtilis* biofilm development in the presence of soil clay minerals and iron oxides. *NPJ Biofilms and Microbiomes, 3*, 1–9.

36. Makusa, G. P. (2013). Soil stabilization methods and materials in engineering practice: State of the art review. *Luleå Tekniska Universitet*.
37. Meyer, F., Bang, S., Min, S., Stetler, L., & Bang, S. (2011). Microbiologically-induced soil stabilization: Application of *Sporosarcina pasteurii* for fugitive dust control. In *Geo-frontiers 2011: Advances in Geotechnical Engineering* (pp. 4002–4011).
38. Pacheco-Torgal, F. (2016). Introduction to biopolymers and biotech admixtures for eco-efficient construction materials. In F. Pacheco-Torgal, V. Ivanov, N. Karak, & H. Jonkers (Eds.), *Biopolymers and Biotech Admixtures for Eco-Efficient Construction Materials* (1st ed., pp. 1–5). England: Woodhead Publishing.
39. Paige-Green, P. (2008). Dealing with road subgrade problems in Southern Africa. In *The 12th International Conference of the International Association for Computer Methods and Advances in Geomechanics (IACMAG)* (pp. 4345–4353). India.
40. Samang, L., Harianto, T., & Djide, M. N. (2018). Bearing capacity improvement of soft soil subgrade layer with bio-stabilized *Bacillus subtilis*. In *MATEC Web of Conferences, EDP Sciences*.
41. Sarda, D., Choonia, H. S., Sarode, D., & Lele, S. (2009). Biocalcification by *Bacillus pasteurii* urease: A novel application. *Journal of Industrial Microbiology & Biotechnology, 36*, 1111–1115.
42. Schallmey, M., Singh, A., & Ward, O. P. (2004). Developments in the use of *Bacillus* species for industrial production. *Canadian Journal of Microbiology, 50*, 1–17. https://doi.org/10.1139/w03-076.
43. Stabnikov, V., Naeimi, M., Ivanov, V., & Chu, J. (2011). Formation of water-impermeable crust on sand surface using biocement. *Cement and Concrete Research, 41*, 1143–1149.
44. Tastan, E. O., Edil, T. B., Benson, C. H., & Aydilek, A. H. (2011). Stabilization of organic soils using fly ash. *Journal of Geotechnical and Geoenvironmental Engineering, 137*, 819–833.
45. Wijeyesekera, D. C. (2014). Overview of innovations in geotechnical engineering. *International Journal of Industrial Engineering, 6*. https://doi.org/10.1016/j.cemconcomp.2014.08.007.
46. Yu, X., Qian, C., Xue, B., & Wang, X. (2015). The influence of standing time and content of the slurry on biosandstone cemented by biological phosphates. *Construction and Building Materials, 82*, 167–172.
47. Zhang, T., Cai, G., & Liu, S. (2017). Application of lignin-based by-product stabilized silty soil in highway subgrade: A field investigation. *Journal of Cleaner Production, 142*, 4243–4257.

CHAPTER 7

POTENTIAL SUITABILITY OF NATURAL LIME AND WASTE CERAMIC DUST IN MODIFYING PROPERTIES OF HIGHLY PLASTIC CLAY

Contents

INTRODUCTION

Expanding soils are frequently the cause of instability in road foundation construction. The primary cause of road failures and damages is frequently the existence of weak subgrades. Over time, the road structure develops excessive and permanent deformation due to the expansive behavior of highly plastic subgrades. Similarly, pavement distresses and premature failure are unavoidably experienced by roads constructed on this kind of subgrade. Expansive soil is prohibited as an appropriate material for road pavement construction due to its low consistency. Therefore, in the conventional road construction trend, weak subgrade soils are completely or partially removed through excavation prior to the start of the actual construction work. However, it is practically impossible to remove inappropriate soil through excavation if the soil mass's thickness extends to a greater depth. Therefore, employing mechanical improvement techniques with any practical stabilizing agent may be a way to get around the problems associated with feasibility.

When clayey soils are stabilized, reactions occur between the negatively charged clay particles and the cation from stabilizers like limestone and waste ceramic dust, which changes the soil's chemical characteristics in addition to its physical ones. The calcium ion (Ca^{2+}) found in limestone and the sodium ion (Na^{2+}) found in ceramic dust are essential for the development of flocculation-agglutination and efficient cation exchange. Lime added to clay starts a two-phase reaction that improves the properties of the clay soil in the short and long terms. The short-term process suppresses the diffuse water surrounding the clay particles by causing an exchange of cations between the ions of the clay particle and the calcium ions of the lime material. The clay soil's particles bind strongly together due to the suppression of pore water. Similar to this, its long-term advantages stem from the pozzolanic reaction, which is created when liberated SiO_2 and/or Al_2O_3 combine in a critical way to form cementing agents like calcium aluminate hydrate (CAH) and calcium silica hydrate (CSH). Thus, the pozzolanic reaction aids in the creation of cementitious materials with a robust bond. Because lime material has a proven drying action, applying lime stabilization speeds up construction in clay soil even during rainy weather (Araújo, 2019).

Numerous investigations were carried out to look into the crucial roles that waste ceramic and limestone play in changing the highly plastic soils' inherent engineering characteristics for use as subgrades. The stabilizer dosage and mix gradation have a significant impact on the degree of engineering property improvement that can be obtained in a stabilized soft soil. Rakhil and Devi claim that waste ceramic can be easily utilized as a stabilizing material, which has been proven to be beneficial in enhancing the geotechnical characteristics of expansive clay soils. The properties of the soil are improved as well as the waste material's disposal problem resolved when ceramic waste is used for weak soil stabilization. According to Binici, ceramics are renowned for having good specific gravity, stiffness, and stiffness. When ceramic waste was added, Nifana and Palanikumar found that there was a noticeable improvement in the soil index properties, unconfined compressive strength (UCS), swelling potential, and Californian Bearing Ratio. Furthermore, it was found that weak subgrades could be economically and safely stabilized with up to 30% ceramic dust. According to Rakhil and Devi, adding 30% of waste ceramic to soft subgrade caused the Californian Bearing Ratio value to increase by 29% and the plasticity index to decrease by 18%. Additionally, the thickness of the pavement was reduced by 45% as a result of the ceramic material being blended with the subgrade; this appears to have saved 31% of the total cost.

Technically speaking, lime is well-known for its quick hydrating qualities, which make it an excellent stabilizing agent for changing the characteristics of problematic soils like clay. The primary source of the strength gain is the chemical reaction that occurs between the minerals in clay and lime. The addition of 5% lime decreased the plasticity index of compressible soil to a level where its compressibility properties could be controlled, according to a study on lime kiln dust for treated subgrades. The study's findings also showed that there was no discernible difference in the subgrade's swelling index improvement levels between the lime dosages of 5% and 8%. Additionally, the results of the research that Prasad et al., It was shown that 4% is the optimal lime dosage to modify the desired properties of the soft subgrade. This was achieved by stabilizing the weak subgrade using 1% to 5% of lime and 0% to 1.5% dosage of waste plastic fiber. In addition, the soil's ideal moisture content decreased as the amount of lime increased.

For both the individual and combined amounts of 3% lime and 15% bagasse ash, Meron and Samuel examined the stabilizing effectiveness of the two materials. The study's findings, in contrast to many earlier studies, showed that a gradual increase in lime content was accompanied by a rise in the magnitude of the ideal moisture content. There have been many studies on the use of lime as a stabilizing material, but little is known about the ability of waste ceramic to stabilize, according to published research. Furthermore, no comparative analysis of the two agents' respective performances in stabilizing problematic subgrades was done. The earlier research primarily examined the individual effects of the two additives; the combined impact of these additions on altering the characteristics of weak soils was not recorded. Therefore, in order to assess the stabilizing potential and performance of waste ceramic dust and lime on highly plastic subgrade, a comparative study was carried out in research. The degree of improvement attained for every parameter was contrasted with the requirements set forth by the Ethiopian Roads Authority for road construction.

7.1. MATERIALS AND METHODS

7.1.1. Characterization of the Materials Used

Three distinct materials were taken into consideration for the experimental analysis in research: limestone, waste ceramic dust, and expansive clay soil. The material regarded as a weak subgrade that is incapable of serving as a pavement foundation is expansive clay. Therefore, it is unlikely that it will be used directly for construction without the use of any improvement mechanisms. Regarding this, different mix proportions are used to use waste ceramic dust and limestone as stabilizing agents. The effects of separately mixing and blending these two materials with the weak subgrade at different mix ratios are examined. In the end, the three components are combined to evaluate how well the additives work together to change the clay subgrade's characteristics.

7.1.1.1. Clay Soil

A sample of the expansive soil that was thought to be improved for the study came from the Asendabo town's periphery in the Jimma Zone of Ethiopia. The soil samples were taken along the Addis Ababa-Jimma trunk road, close to the damaged road segment that connected Asendabo town with Deneba town, at a depth of two meters below the natural ground surface. As a result, the samples that were gathered belong to the road subgrade and share the same physical characteristics as the damaged pavement's foundation.

Expansive soils have distinct qualities and qualities that are greatly influenced by the minerals that make up the soil. Table 7.1 lists a few research articles. Clay soils are invariably moisture-sensitive and greatly impacted by variations in water content; as a result, they absorb a lot of water and quickly swell and lose strength.

These expansive soils have a propensity to heave during wet conditions and shrink in volume during dry seasons. They are also easily compressible when wet. Broad soils are also well-known for having a lot of clay minerals in them, which allow them to absorb a lot of water after rain. Highly expansive soils are compressible soils with a free swell index greater than 50%. These soils experience volumetric changes that cause distortion in pavements (Bhengu and Allopi, 2017).

Table 7.1. *Research Methods Regarding the Failure Mechanism of Expansive Soil Slopes*

Research Methods	**Advantages**	**Disadvantages**
Field test	1. High accuracy and reliability of results. 2. In-situ (undisturbed) test soil. 3. Test environment close to the actual project site conditions.	1. Long cycle; non-repeatable. 2. High cost. 3. Difficulty in controlling the test conditions, with many constraints.
Centrifugal model test	1. The stress and strain of the model and the prototype are equal, with similar deformation. 2. The use of an artificial gravity field can compensate for the loss of stress in the test and reproduce the characteristics of the prototype in a short time. 3. The small size of the model saves labor and material resources. 4. Test conditions can be controlled.	1. High equipment cost. 2. High requirements for material selection. 3. Particle size effect and boundary effect have an impact.
Physical model test	1. Economical and intuitive. 2. Relatively low equipment requirements. 3. Test conditions can be controlled; not subject to the limitations and influence of environmental conditions.	1. The stress level of the model is much lower than that of the prototype, making it difficult to reproduce the characteristics of the prototype. 2. The test soil is remodeled soil (Al-Sharif and Attom, 2014). 3. Boundary effects have an impact.

Numerical simulation	1. Low cost and short time. 2. Easy to operate and repeatable. 3. Can be used in combination with field tests and model tests. 4. It is easy to analyze the stability of expansive soil slopes under different operating conditions by changing the variables (Kassahun, 2018). 5. Can be used with access to various parameters of the slope.	1. Need to make more simplifications and assumptions, resulting in a certain degree of error. 2. Incorrect parameters, conditions, and assumptions can lead to inaccuracies.

Source: Luo P, Ma M. (2024). Failure mechanisms and protection measures for expansive soil slopes: A review. Sustainability, 16(12):5127. https://doi.org/10.3390/su16125127.

As can be seen from the research, silicon oxide (SiO_2) and aluminum oxide (Al_2O_3), which account for 60.8% and 21.26% of the soil's weight, respectively, are the two most common chemicals. On the other hand, the least important chemicals in terms of composition are water (H_2O), phosphorus pent oxide (P_2O_5), and manganese oxide (MnO). As a result, the typical characteristics of silicon dioxide, the dominant oxide, have a significant impact on the intrinsic properties of the expansive soil (Figure 7.1).

Figure 7.1. *The failure of expansive soil slopes with different protection methods: (a) no protection; (b) concrete protection; (c) concrete frame and vegetation protection; and (d) vegetation protection.*

Source: Luo P, Ma M. (2024). Failure mechanisms and protection measures for expansive soil slopes: A review. Sustainability, 16(12):5127. https://doi.org/10.3390/su16125127.

Texturally, the weak soil is mostly made up of fine particles with very little sand particles present. Because of this, the majority of the soil is made up of fine grains (clay and silt) with particle sizes smaller than those of many 200 sieve. There is absolutely no gravel soil fraction. The distribution of soil particle sizes shows that nearly 94.2. The sand fraction makes up 61.8% of the soil mass, and 0% of the particles pass the No. 200 sieve. According to its type, the soil is prohibited as having poor engineering properties and should not be used directly for subgrade construction without using stabilization of some kind. Furthermore, the soil's consideration for road construction is limited due to its poorly graded (uniformly graded) type and poorly proportioned constituent particles (Figure 7.2).

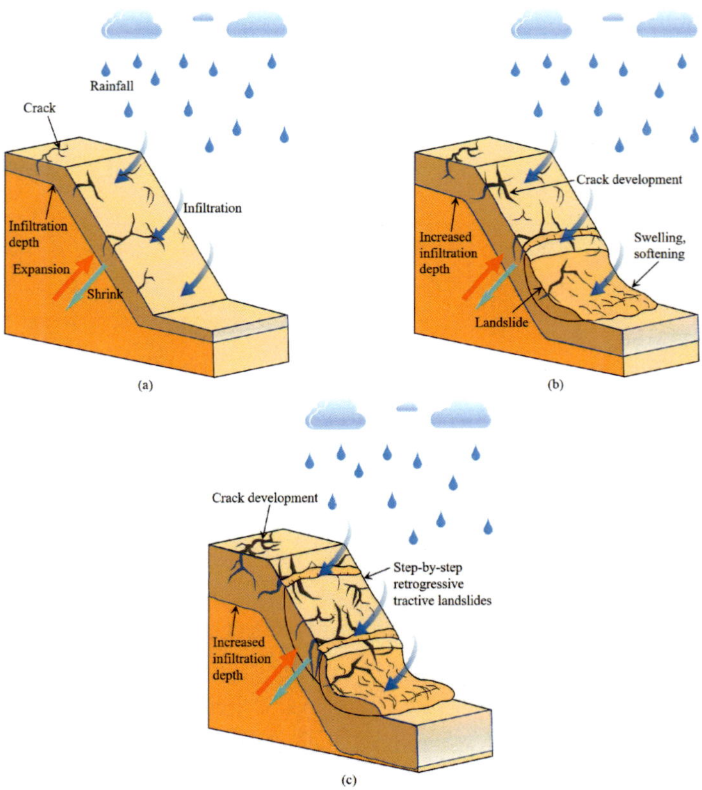

Figure 7.2. *Destabilization and failure of expansive soil slopes: (a) initial state; (b) experience of repeated wet–dry cycles; and (c) final failure.*

Source: Luo P, Ma M. (2024). Failure mechanisms and protection measures for expansive soil slopes: A review. Sustainability.; 16(12):5127. https://doi.org/10.3390/su16125127.

7.1.1.2. Waste Ceramic Dust (WCD)

Ceramic factories and locally accessible sources provided the ceramic dust. The direct stabilization of industrial waste has helped to lower pollution levels in the environment. The issue of how to dispose of leftover ceramic material is a major environmental concern in many places.

Table 7.2. The Effect of Different Factors on the Stability of Expansive Soil Slopes

Factor Properties	Influencing Factors	Effects on Slopes
External factors	Rainfall	1. Soil swelling and softening. 2. Pour water pressure increases and effective stress decreases (Amadi and Okeiyi, 2017). 3. Soil shear strength decreases (Amadi and Okeiyi, 2017). 4. Soil self-weight increases, sliding force increases. 5. Slope erosion damage.
	Wet–dry cycles	1. Increased crack development destroying the integrity of the soil. 2. Significant changes in soil water content. 3. Generate cumulative deformation. 4. Reduction of the slope safety factor (Maureen, 2009).
	Human activities	1. Excavation unloading (Mutaz, and Muawia, 2014; Oleng, 2018). 2. Loading on the top of the slope, accelerating instability.
Internal factors	Cracks	1. Provision of a direct pathway for rainwater infiltration. 2. Reduction in soil strength. 3. Tends to the formation of weak or sliding surfaces. 4. Acceleration of soil erosion.
	Swelling–shrinkage	1. Production of uneven deformation and acceleration of crack development. 2. Causes uneven soil stresses inside the slope; production of stress concentration; triggering factor of strength failure.

	Soil properties	1. Increased expansion rate increases the potential for destabilization. 2. Increased hydrophilic mineral content of the soil increases the potential for induced cracking (Prasad et al. (2018). 3. Decreased soil density and increased degree of weathering increase the likelihood of erosion damage. 4. Increased degree of over-consolidation decreases the safety factor of expansive soil slopes.

Source: Luo P, Ma M. (2024). Failure mechanisms and protection measures for expansive soil slopes: A review. Sustainability, 16(12):5127. https://doi.org/10.3390/su16125127.

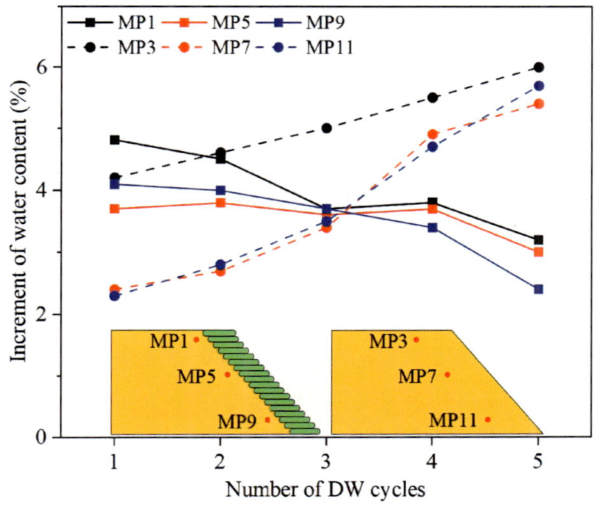

Figure 7.3. *Effect of soil bag stabilization on the change of water content of expansive soil slopes under wet–dry cycles.*

Source: Luo P, Ma M. (2024). Failure mechanisms and protection measures for expansive soil slopes: A review. Sustainability, 16(12):5127. https://doi.org/10.3390/su16125127.

Because of this, it is crucial to use it as an admixture to stabilize weak ground in order to reduce pollution and increase strength. Larger values of mix proportions are often needed to achieve the desired level of improvement.

However, in some real-world scenarios, a small amount of ceramic blend—up to 5%—might stand out to enhance the compressible subgrade soils' strength and capacity to support loads. Because waste ceramic product has excellent characteristics like high strength, long service life, chemical inertness, non-toxicity, and specific porosity, it is frequently used to stabilize weak and compressible soils (Jalali, 2010) (Table 7.2; Figure 7.3).

7.1.1.3. Natural Lime (Limestone)

The naturally occurring lime sources close to Ambo town provided the limestone used as the stabilizing agent in the study. Lime, whether processed or unprocessed, can be applied to weak soils to increase their capacity to support loads and their plasticity. Without any processing, the limestone sample used in this investigation was taken straight from easily accessible deposits (limestone rock mass). The primary chemical component of limestone is calcium carbonate. Lime is an inorganic material that mostly consists of carbonates, oxides, and hydroxides and contains calcium. Strictly speaking, Limestone is calcium carbonate and lime is calcium oxide or hydroxide., but naturally occurring limes are mostly fossiliferous and contain a lot of stylolite. Good quality lime has the following qualities that make it an excellent choice for an engineering material: it is easily workable, has good plasticity, resists moisture well, stiffens quickly, stabilizes soils, and exhibits excellent cementation behavior. Lime's quick dehydration properties make it an excellent choice for amending fine-grained soils, which is another wonderful benefit of using lime for weak soil stabilization. In contrast to ceramic dust, a much smaller dosage of natural lime is needed to meet road specifications and improve weak soils to their best potential. Although the amount of mix required depends largely on the inherent qualities of weak soils, compressible soil can be effectively modified with a small amount of lime (4–7%) (Table 7.3).

Table 7.3. Groups of Limestones/Correctives Product Samples Based on the Efficiency Neutralizing Power and MgO and CaO Contents

Samples	**ENP/MgO Contents**		**2 mm Mesh**	**Retained 0.84 mm Mesh**	**0.3 mm Mesh**	**NP (%)**	**ENP (%)**	**MgO (%)**	**CaO (%)**	**Ca/Mg Ratio**
1	70–80	high Mg	0	14.4	28.7	102.03	78.56	10.12	31.43	3.7

2			0	19.1	30.6	104.23	75.55	10.16	28.16	3.3
3			0	19.4	35.0	103.19	72.73	10.12	28.65	3.4
4	80–99	high Mg	0	6.5	13.0	108.88	97.56	10.09	29.28	3.4
5			0	12.0	22.5	111.44	90.71	10.06	30.08	3.5
6			0	7.0	15.3	110.51	97.56	10.45	29.99	3.4
7		low Mg	0	7.2	24.5	104.24	88.02	5.51	43.2	9.3
8			0	9.2	25.6	100.29	82.64	6.55	37.92	6.9
9			0	8.6	20.4	104.35	88.66	7.01	45.16	7.6
10	>99	high Mg	0	0	0	99.47	99.47	10.32	30.53	3.5
11			0	0	0	159.89	159.89	10.59	30.62	3.4
12			0	0	0	155.94	155.94	10.08	38.64	4.5
13		low Mg	0	0	0	105.37	105.37	0.61	43.1	83.7
14			0	0	0	99.13	99.13	0.56	45.92	97.1
15			0	0	0	107.72	107.72	0.51	56.72	131.7
min	$n = 52$						69.2	0.5	23.0	
mean	$n = 52$						88.2	5.6	31.6	
max	$n = 52$						159.9	10.6	56.7	

Efficiency neutralizing power (ENP).

Source: Bortoluzzi EC, Garibotti A, Tiecher T, dos Santos DR, Moterle DF, Fiorin JE. (2022). Revisiting limestone quality for soil liming purpose. Minerals, 12(12):1522. https://doi.org/10.3390/min12121522.

The chemical characteristics of the natural limestone are depicted in Figure 7.4. As can be seen in the figure, the main chemical present in limestone is calcium oxide (CaO), which is followed by loss on ignition (LOI) and silicon oxide (SiO_2), which account for 44.4%, 24.8%, and 10.83% of the weight, respectively. In connection with this, the rare substances present in limestone are titanium oxide (TiO_2), manganese oxide (MnO), sodium oxide (Na_2O), and phosphorus pent oxide (P_2O_5).

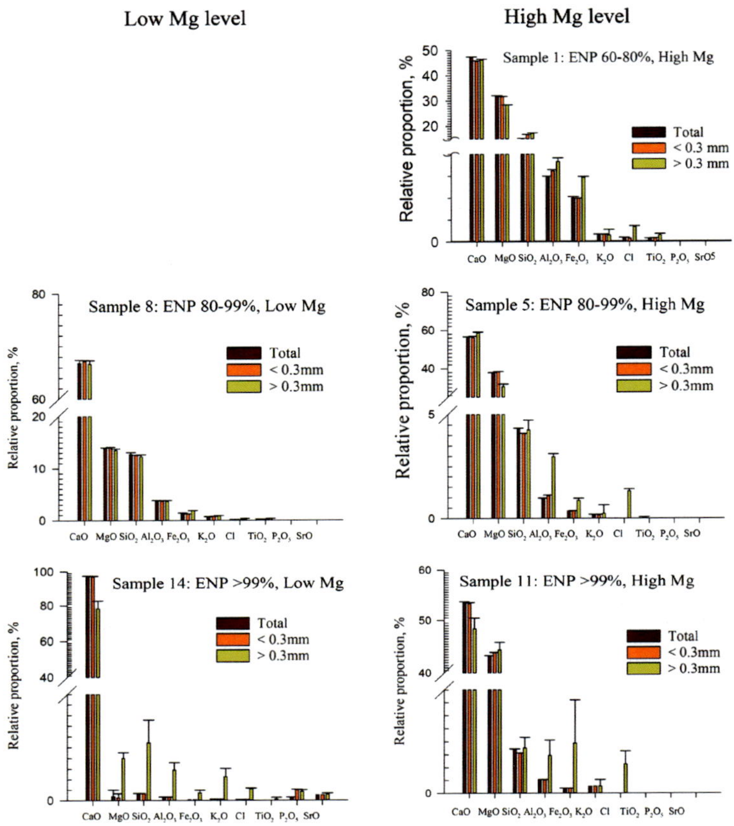

Figure 7.4. *Chemical composition using X-ray fluorescence by dispersive energy from representative lime/corrective samples separated into three groups of efficiency neutralizing power—ENP and two Mg levels (low and high).*

Source: Bortoluzzi EC, Garibotti A, Tiecher T, dos Santos DR, Moterle DF, Fiorin JE. (2022). Revisiting limestone quality for soil liming purpose. Minerals, 12(12):1522. https://doi.org/10.3390/min12121522.

7.1.2. The Experimental Tests

7.1.2.1. Sample Preparation and Blending

Using hand excavation tools, test pits were dug up to a depth of two meters, and samples of the weak and compressible clay were removed. The samples were carefully transported to the soil laboratory and stored in sampling bags to prevent substantial moisture loss. The natural limestone mass was manually crushed into finer pieces using a 425 μm sieve, and the experimental tests were conducted in accordance with AASHTO M-145 recommendations. In order to examine the impact of mix ratio variations on the performance of the weak subgrade material, five distinct mix proportions of natural lime, ranging from 2 to 10%, were taken into consideration. In order to evenly distribute the stabilizer, the two materials were combined using a mechanical mixer.

Similarly, to make mixing easier and more consistent with the weak soil, blocks of ceramic waste from nearby factories were crushed to the same size as limestone. To ensure that the mix contained uniformly sized particles of varying sizes, the ceramic was manually crushed with a hammer. Similarly, the crushed ceramic waste is made up of particles with different diameters in order to deliberately prevent some particles with the same size from predominating. As a result, eight different mix ratios, ranging from 2 to 25%, were used to combine the weak subgrade material and ceramic waste. Following that, a mixture of the materials was used directly for various laboratory tests to ascertain the swelling index, moisture content, unit weight, Atterberg limits, compaction characteristics, and California bearing ratio.

7.1.2.2. The Test Design and Considered Case Scenarios

Four distinct case scenarios are included in the overall test procedure of the planned experimental test for comparative analysis. In the first case, the only weak subgrade material is characterized without any stabilizing agent blended in. In this instance, the material's suitability and workability were assessed by determining the essential physical characteristics of the weak and compressible subgrade. In the second case, the critical effect of limestone is examined through dosage variation. Therefore, for each parameter that has been determined, a potential clear relationship between the variation of the mix ratio and the achieved degree of improvement can be established. Thirdly, in order to assess the stabilizing effectiveness of waste ceramic dust (WCD) independently of natural lime, a combination of crushed waste ceramic and weak subgrade was employed without the addition of any additional ingredients.

In the fourth case, a combination of the three materials—waste ceramic, limestone, and weak soil—was used to examine the combined impact of the two materials on the subgrade properties. Several mix ratio combinations of the leftover ceramic and limestone were used to achieve this. Similarly, five variants with various additive amounts were taken into account. Around 2% natural lime plus 15% WCD, 4% natural lime plus 15% WCD, 6% natural lime plus 15% WCD, 8% natural lime plus 15% WCD, and 10% natural lime plus 15% WCD are these combinations. As a result, the stabilizing agents' individual and combined improvement potential was carefully considered (Table 7.4).

Table 7.4. Mineral Composition and Quantification Based on the X-Ray Diffraction (XRD) Patterns from Limestone/Corrective Powder Samples

Samples	Mineral	Ideal Chemical Formula	XRD Peaks (d = nm) Intensity (I/Io)	Relative Proportion Percent (%)
(1), 2, 3	Ankerite	$CaFe_2 + 0.6Mg_{0.3}Mn_2 + 0.1(CO_3)_2$	0.289 (1), 0.181 (0.06), 0.219 (0.06)	44.1
	Dolomite	$CaMg(CO_3)_2$	0.288 (1), 0.178 (0.6), 0.219 (0.5)	37.6
	Sillimanite	$(Al_2O_3)(SiO_2)$	0.342 (1), 0.337 (0.65), 0.22 (0.6)	8.9
	Otavite	$Cd(CO_3)$	0.295 (1), 0.378 (0.8), 0.246 (0.35)	3.8
	Calcite Mica	$CaCO_3$	0.303 (1), 0.209 (0.18), 0.228 (0.18) 1.00	1.3~
4, (5), 6	Dolomite	$CaMg(CO_3)_2$	0.288 (1), 0.178 (0.6), 0.219 (0.5)	84.5
	Ankerite	$CaFe_2 + 0.6Mg_{0.3}Mn_2 + 0.1(CO_3)_2$	0.289 (1), 0.181 (0.06), 0.219 (0.06)	7.4
	Otavite	$Cd(CO_3)$	0.295 (1), 0.378 (0.8), 0.246 (0.35)	5.8
	Quartzo	SiO_2		0.6
7, (8), 9	Calcite	$CaCO_3$	0.303 (1), 0.209 (0.18), 0.228 (0.18)	54.0

	Dolomite	$CaMg(CO_3)_2$	0.288 (1), 0.178 (0.6), 0.219 (0.5)	36.1
	Quartzo	SiO_2		8.7
	Otavite	$Cd(CO_3)$	0.295 (1), 0.378 (0.8), 0.246 (0.35)	1.2
	Mica		1.00	~
10, (11), 12	Calcite	$CaCO_3$	0.303 (1), 0.209 (0.18), 0.228 (0.18)	66.3
	Portland-ite	$Ca(OH)_2$	0.262 (1), 0.49 (0.74), 0.192 (0.42)	14.3
	Periclase	MgO	0.210 (1), 0.148 (0.52), 0.122 (0.12)	13.8
13, (14), 15	Calcite	$CaCO_3$	0.303 (1), 0.209 (0.18), 0.228 (0.18)	87.1
	Briartite	$Cu_2Zn_{0.75}Fe^{2+}{}_{0.25}GeS_4$	0.306 (1), 0.187 (0.5), 0.189 (0.5)	1.2

Relative proportion among mineral assemblage was obtained in the Diffrac-EVA software.

Source: Bortoluzzi EC, Garibotti A, Tiecher T, dos Santos DR, Moterle DF, Fiorin JE. (2022). Revisiting limestone quality for soil liming purpose. Minerals, 12(12):1522. https://doi.org/10.3390/min12121522.

The Ethiopian Road Authority's specifications for road construction were then compared with the improvement performance of the final three cases (Cases two through four). A thorough investigation into the individual effects of these stabilizing agents makes it possible to conduct an efficient comparative analysis, particularly from the perspectives of economy and workability. Ultimately, out of the three cases, the most effective and workable option may have been chosen for use in subsequent subgrade stabilization procedures. The stabilizer dosages taken into consideration in this study were derived in part from previously published literature as well as from practical trends.

7.1.2.3. The Subgrade Tests and Procedures

Moisture content, compaction characteristics, Atterberg limits (plastic limit, liquid limit, and plastic index), California bearing ratio (CBR), specific gravity, swelling index, and soil particle distribution are among the soil parameters taken into account in research to achieve the research objective. Before all, the AASHTO T-265 standard test method for laboratory determination of soil water content was used to determine the moisture content of the naturally weak soil. The maximum dry density (MDD) and ideal moisture content of the natural and treated subgrades were ascertained through the standard compaction test. Raising the soil's density would enhance many of its engineering qualities, including its strength, stiffness, resistance to shrinkage, and imperviousness. The test was carried out in compliance with AASHTO T-99 guidelines. This technique is used for the soil particles that are held on the No. 4 (4.75 mm) sieve.

Another significant soil property that is used as a criterion for improvement is the CBR. It actually aims to determine the relationship between force and penetration by expressing the force applied by a plunger and the depth of penetration into the specimen. The shearing resistance of soil under regulated moisture and density conditions is indirectly measured by the CBR test. It establishes the subgrade soil's strength and capacity to swell. The test was carried out as per (AASHTO T-193-93) test procedure. Furthermore, Atterberg limits (Liquid limit, plastic limit and plastic index) were determined for air-dried samples by using Casagrande apparatus. The test was conducted using the standard reference provided by AASHTO T-89 and 90. For the treated soil with varied amounts of limestone and waste ceramic dust, the same process was used. Together with the previously listed tests, a specific gravity test was carried out utilizing a water pyrometer in accordance with AASHTO T 100-93 standard. It is a measurement of the soil's density in relation to the water's density. The most basic yet crucial test used to determine the soil's capacity to swell is the free swell test. The increase in soil volume upon submersion in water, free from external constraints, is known as the free swell index (Al-Sharif and Attom, 2014).

Grain size analysis was performed with respect to the particle size distribution using the ASTM D422-63 test procedure. For coarse-grained soils, a set of sieves was utilized for mechanical analysis, and for fine-grained soils, a hydrometer. For soils with both coarser and finer sizes, testing was done using both mechanical and hydrometer techniques. Finally, the natural weak soil was classified using the AASHTO and unified soil classification systems (USCSs).

7.1.2.4. Chemical Analysis

The materials taken into consideration for the study were characterized through chemical analysis. To ascertain the chemical composition of the waste ceramic dust, limestone, and clay soil, an X-ray diffraction (XRD) test was performed. The primary purpose of X-ray diffraction analysis is to identify the mineralogical phases that are present in various materials. Samples of the three materials were dried in a representative oven and then ground into a powder that meets the No. 200 mesh (0.07 mm opening). The powder samples underwent continuous scanning speed and a step scan with a 1 s time step from 10° to 75° (2θ). The voltage was set to 30 kV with 25 mA for testing and analysis, and the scanning speed for the XRD test was 0.02°/sec. Diffraction is a phenomenon that happens when a wave runs into a gap or obstruction. It is the bending of waves around an obstacle's corners or through an opening into the obstacle's geometric shadow. Essentially, the average spacing and orientation between layers and crystals are measured using X-ray diffraction. Additionally, research is done on the dimensions, morphology, internal stress, and crystal structure of tiny crystalline regions.

7.1.3. Clay Soil Stabilization and Chemical Reactions

When added to expansive clay soil, stabilizing agents are crucial to the process of enhancing the soil's qualities in three ways. These three roles are stabilization, modification, and drying. The chemical reaction between the water and the additives causes the clay soil's moisture content to rapidly fall during the first phase. Secondly, the incorporation of stabilizing agents diminishes the plasticity and minimizes the characteristics of swelling and shrinking. In connection with this, adding lime to clay soil lowers the MDD of the clay and raises the ideal moisture content. Following the addition of lime, the soil undergoes the aforementioned changes in properties quickly. The third stage of the improvement process focuses on making a long-term difference to the clay soil's characteristics. Long-term strength, a steady decrease in plasticity, and a permanent decrease in shrinkage, swelling, and resistance to thawing and extended soaking are some of these enhancements.

When lime material is added to clay soils with high moisture content, a number of reactions occur that improve the clay soil's characteristics. There are primarily two types of these chemical reactions: pozzolanic and colloidal. The cation exchange, flocculation, and carbonation are a few of the immediate (colloidal) reactions. The dissolution of clay particle alumina and silica sheets causes the pozzolanic reaction to occur eventually. These reactions can happen right away or over an extended period of time after mixing the additives with clay soils. Carbonation, in contrast to pozzolanic and other colloidal reactions, is an undesirable reaction that impairs the stabilization process. In clay-lime mixtures, this reaction occurs. Instead of the significant cementitious materials like calcium silicate hydrate (CSH) and calcium aluminate hydrate (CAH), calcium carbonate is formed in this reaction between lime material and carbon dioxide.

7.1.3.1. Cation Exchange

One of the key processes that primarily affects how soil's plasticity characteristics change is cation exchange. Positively charged ions, known as cations, include potassium (K^+), sodium (Na^+), calcium (Ca^{2+}), manganese (Mn^{2+}), zinc (Zn^{2+}), copper (Cu^{2+}), aluminum (Al^{3+}), iron (Fe^{2+}), manganese (Mn^{2+}), and sodium (Na+). Because clay particles typically have a negative charge, clay soils have a cation exchange capacity. The entire soil is electrically neutral. Therefore, the positive charge of the cations in the limestone and leftover ceramic dust balances the negative charge of the clay particles. The cation exchange capacity value of fine-textured soils, such as clay, is higher than that of coarse-grained soils. Cation exchange capacity (CEC) refers to the soil's capacity to retain cations. The total number of ions per unit weight of soil is another definition of the cation exchange capacity of soil. Milliequivalents per 100 grams of soil are the unit of measurement.

In soil, cations are exchanged according to their capacity for replacement. Higher valence cations take the place of lower valence cations. Larger cations take the place of smaller ones when the cations have the same valence. For example, the valence of calcium (Ca^{2+}) and magnesium (Mg^{2+}) is the same. But because the valence of the latter cation is greater than that of the former, Mg^{2+} is replaced earlier than Ca^{2+}. The weaker cations in the expansive clay soil are replaced by the calcium cation from the lime when an appropriate amount of lime material is added. The process of cation exchange causes the diffuse water layer that surrounds the clay particles to thin, bringing the particles closer together and promoting the formation of flocculation. Many elements, including soil texture, the type of clay minerals present, PH level, and organic content, have an impact on the cation exchange of the soil. The positively charged cations carried by the limestone and ceramic material are attracted to and retained by the negatively charged clay colloids. Higher clay-content soils have a greater capacity for cation exchange than lower clay-content soils. The pH of the water between soil particles is maintained at a higher level due to the formation of calcium hydroxide, which is the result of a reaction between quick and water. Consequently, the clay particles' silicates (SiO_2) and aluminates (Al_2O_3) dissolve. As the pore water's pH rises, the silicates become ionized, increasing the cation exchange capacity (Araújo, 2019).

7.1.3.2. Flocculation-Agglomeration

The combined effect of flocculation-agglomeration and cation exchange processes is clay soil stabilization. With time, the plastic soil texture transforms into non-plastic materials due to these two processes, which also increase workability, increase interparticle friction, and decrease the adsorbed water layer. The process known as flocculation-agglomeration causes a change in the texture of the soil as negatively charged clay particles unite to form large-sized particles. This occurs when the cation exchange reduces the amount of pore water that is present between the clay particles. Around 45% of the limestone sample used for chemical analysis is made up of calcium oxide. The lime material exchanges its calcium ions (Ca^{++}) for the sodium ions (Na^{+}) carried by the ceramic material. Consequently, the plastic clay soil acquires a flocculated structure rather than a dispersed structural arrangement, thereby enhancing its overall workability.

7.1.3.3. Pozzolanic Reaction

Strong cementitious material is permanently created when hydrated lime reacts with clay soil particles. The chemical reaction that occurs between quicklime and water yields hydrated lime. Lime material with a high calcium content and no more than 5% magnesium content is recommended for the efficient treatment of clay soil. When lime is added to clay soils, the calcium ions are first absorbed by the clay minerals or combined with them, which greatly reduces the workability of the compressible clay soil. Pozzolans are essential to the long-term lime stabilization process, particularly in the creation of cementitious bonds. Pozzolanic materials are well-known for their ability to cement, which results in strong bonding. Their reaction and action in expansive clay soils build up firmly bonded particles because of their cementation and bonding characteristics, which directly contribute to reducing the clay soil's tendency to swell and improving its strength properties.

Clay soils are naturally home to these pozzolans. Calcium-silicate-hydrates (C-S-H) and calcium-aluminate-hydrates (C-A-H) are the products of the reactions between pozzolans like silica and alumina, calcium present in lime, and water. Although silicon and aluminum are common to clay minerals, their proportions differ due to their unique structures. When lime is added to clay in the presence of water, pozzolanic reactions involving clay soil minerals and lime occur. Calcium silicate hydrate (CSH–$3CaO_2SiO_2$ $3H_2O$) and calcium aluminate hydrate (CAH—$3CaOAl_2O_3$ $Ca(OH)_2$ $12H_2O$) are cementing agents that change the structure of the soil, lessen its plasticity, and increase its workability when combined with liberated silicon dioxide (SiO_2) and/or aluminum oxide (Al_2O_3).

Clay soil's pH, organic carbon content, carbonate content, degree of weathering, natural drainage, exchangeable sodium content, clay mineralogy, extractable iron, and silica-alumina ratio are some of the factors that affect lime's reactivity.

7.1.4. Specifications and Standards Used

The current and relevant road construction specifications were compared with the stabilizing potential of the additives and their combination. As a result, the degrees of improvement were compared to the Indian standard and the 2013 road construction specification published by the Ethiopian Roads Authority (ERA). The improved soil's Liquid Limit (LL), Plastic Index (PI), and CBR were compared to the ERA 2013 specification, while the soil's swelling potential was assessed using the Indian Standard specification. The 2013 Ethiopian Roads Authority (ERA) Specification for building new roads places a strong focus on specific soil parameters, including the CBR, Plastic Index, Liquid Limit, Particle Size Distribution, and Compaction Degree (Table 7.5).

Table 7.5. The ERA 2013 and IS Specifications for Road Construction (Subgrade)

Sl. No.	Subgrade Properties	Recommendation of ERA 2013 Specification	IS: 2911 Part III- 1980
1	Liquid Limit (PL) (%)	<30	
2	Plastic Limit (LL) (%)	<60	
3	California Bearing Capacity (CBR)	>3	
4	CBR Swell (%)	<2	
5	Free Swell (%)	–	<20

Source: https://www.ncbi.nlm.nih.gov/pmc/articles/PMC9582719/table/tbl5/?report=objectonly.

7.2. EFFECTS OF NATURAL LIME AND WASTE CERAMIC DUST IN MODIFYING PROPERTIES OF HIGHLY PLASTIC CLAY

7.2.1. Effect of Natural Lime and WCD on Compaction Characteristics of the Clay

The study's consideration of additives had an impact on the expansive clay subgrade's compaction properties through their blend. It was shown that the compaction characteristics of the clay subgrade are significantly impacted by variations in the stabilizer dosage. The expansive clay's MDD increased almost linearly with a gradual increase in the dosage of waste ceramic dust (WCD), but the MDD was sharply decreased with an increase in the amount of natural lime. Additionally, the MDD was increased by 12.4% and 30.6%, respectively, with the addition of 10.30% WCD.

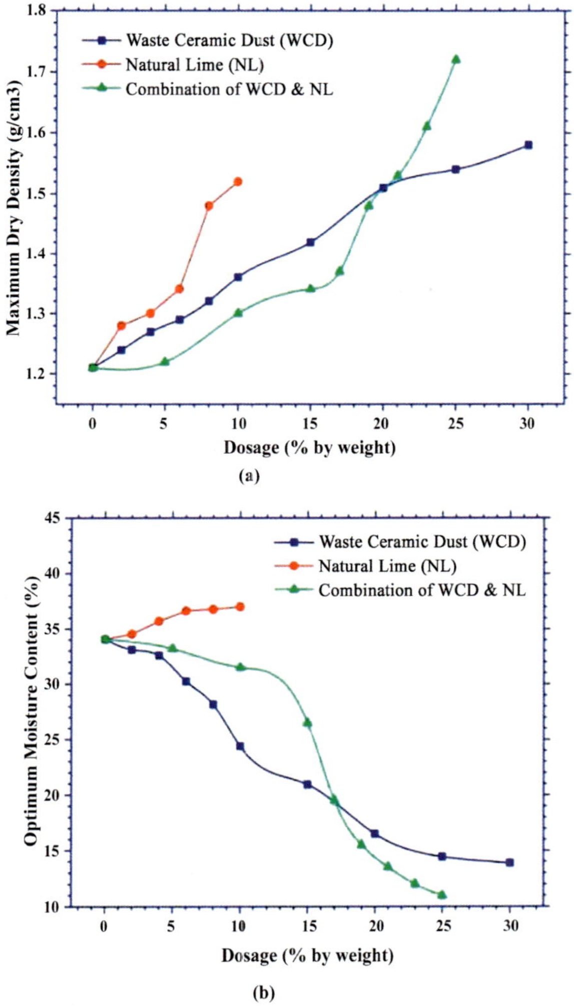

Figure 7.5. *Liming for soil.*

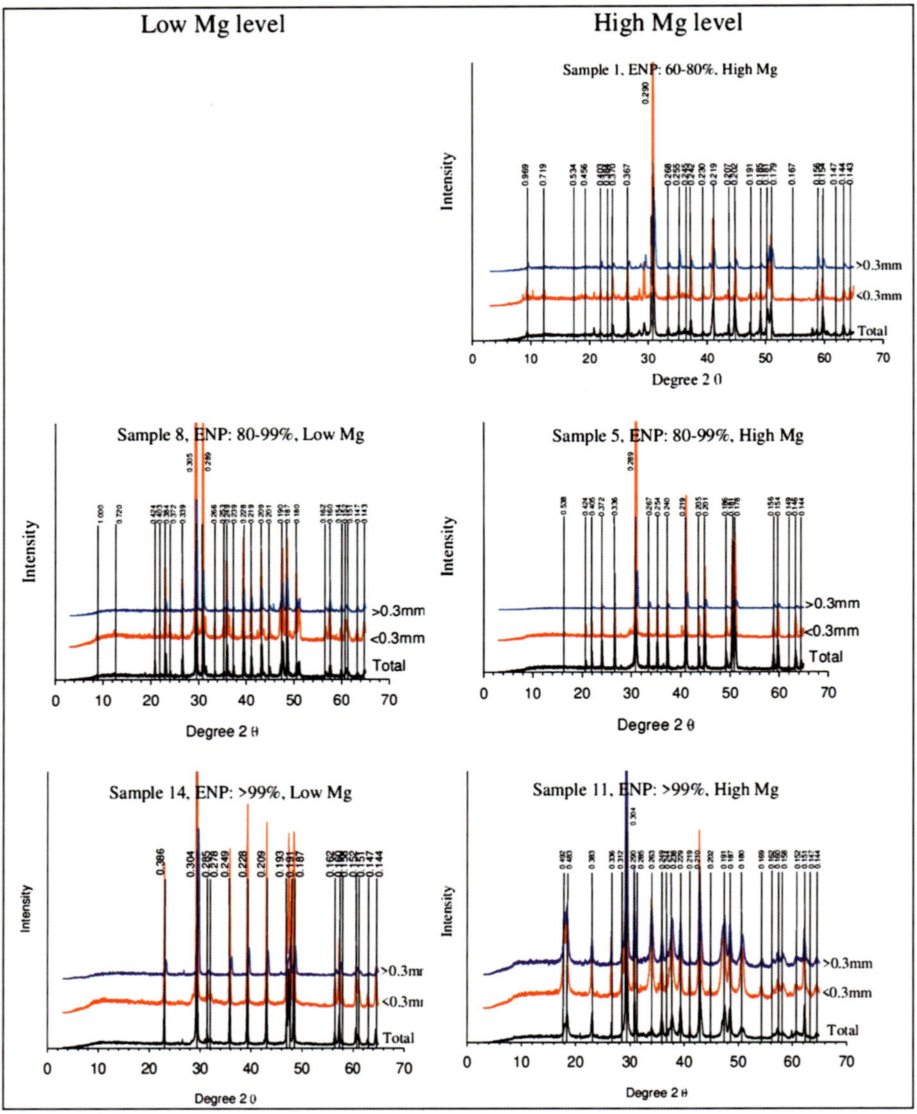

Figure 7.6. X-ray diffraction patterns from representative liming products categorized by efficiency neutralizing power (ENP) and magnesium (Mg) levels (low and high).

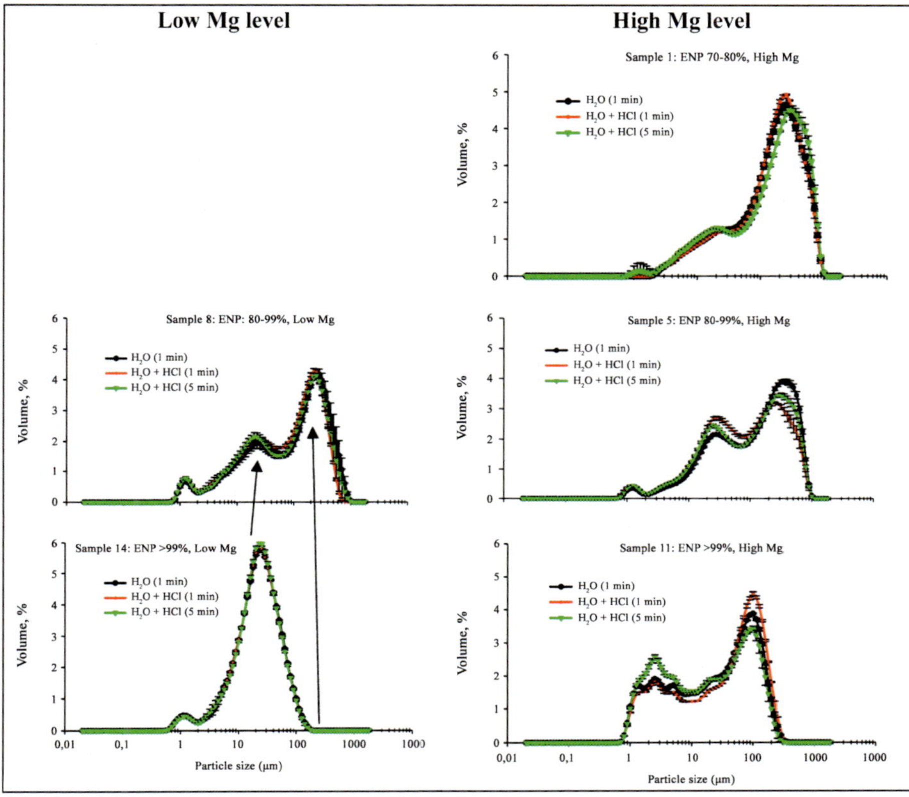

Figure 7.7. Laser diffraction granulometric profiles of lime/corrective samples classified by ENP and magnesium (Mg) levels (low and high).

Principal component analyses considered liming product samples, including their number, mineralogy, and physical and chemical properties. The data set used comprised 15 samples (n = 15). The principal components (i.e., axes PC1 and PC2) explain the magnitudes of percentage data variability. D10 represents the percentage of particle size smaller than 10%. Size fractions of <0.30 mm and 0.3–0.84 mm were evaluated. RE refers to reactivity efficiency, and ENP stands for efficiency neutralizing power. When it comes to the ideal moisture content, the value increases when waste ceramic dust is added, but it decreases when more natural lime is added. The OMC decreased from 34.06% to 24.38% with the addition of 10% WCD. Similarly, adding 30% of waste ceramic dust dosage decreased the ideal moisture content by 44.6%. The decrease in the ideal moisture content may be the result of water being drawn to the fine, dry particles of broken ceramic, which eventually causes the mix's water content to drop. On the other hand, the addition of 10% natural lime increased the soil's OMC by 25.84%, from 34.06% to 42.86%.

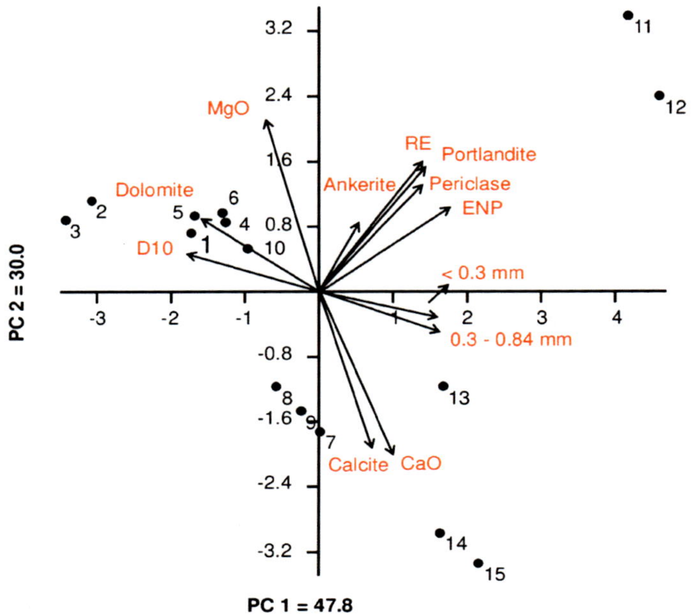

Figure 7.8. Principal component analysis of liming product samples considering their mineralogy, physical properties, and chemical attributes.

Source: Bortoluzzi EC, Garibotti A, Tiecher T, dos Santos DR, Moterle DF, Fiorin JE. (2022). Revisiting limestone quality for soil liming purpose. Minerals, 12(12):1522. https://doi.org/10.3390/min12121522.

When the quantity of natural lime fluctuated more quickly than the waste ceramic dust, the OMC changed in magnitude more quickly. Therefore, compared to waste ceramic dust, a small variation in the amount of natural lime causes a significant change in the OMC of the expansive subgrade soil. According to the results of a study done to look into how cement and lime affected the geotechnical characteristics of clay soil, both the MDD and the ideal moisture content increased as the lime content increased from 2% to 10%. But Arvind K. along with others found that as the amount of lime in the treated clay increases, its OMC rises while its MDD falls.

The interaction between waste ceramic dust and natural lime on the MDD and optimal moisture content of the clay soil is depicted by the green line in Figure 7.5(a) and (b). According to the graphs, the MDD increases as the amount of limestone increases, maintaining a consistent level of waste ceramic dust (15%). Conversely, the optimal moisture content of the clay significantly decreases. The combined effect of 4% natural lime and 15% waste ceramic dust (WCD) in reducing the optimal moisture content is nearly equal to the impact of using 30% WCD independently. The optimal moisture content of the clay subgrade increases noticeably when natural lime is added and its dosage is increased, in the scenario where natural lime is used alone. However, when used in combination with a specific amount of waste ceramic dust, the resulting optimal moisture content decreases. Researchers Rakhil and Devi, as well as Nifana and Palanikumar, suggest that the MDD (MDD) increases while the optimal moisture content (OMC) decreases with an increasing content of WCD (Rakhil and Devi, 2016).

7.2.2. Effect of Natural Lime and Waste Ceramic Dust on CBR and CBR Swell of the Clay Soil

The impact of natural lime dosage and waste ceramic dust on the clay soil's CBR is shown in Figure 7.6(a). As illustrated in the figure, the uncured CBR value increased for all mixes, with the natural lime percentage increasing from 2% to 10% and the waste ceramic dust percentage increasing from 2% to 30%. Similarly, the separate application of 6% WCD and natural lime increased the magnitude of the CBR by 54.73% and 87.24%, respectively. The results of the experimental test showed that, in terms of altering the soil's CBR, 3% natural lime functions comparably to 10% WCD. Therefore, it can be concluded that the expansive soil's CBR is more susceptible to changes in limestone dosage than waste ceramic dust. Clay soil does not become completely stronger in a short amount of time, and this strength gain is directly related to pozzolanic reactions. The development of cementitious materials like CSH and CAH is what gives clayey soils their strength and load-bearing capacity. The strength characteristics of expansive clay soil appear to be enhanced by a gradual increase in the dosage of lime material. However, adding more lime than is necessary eventually weakens the material because too much silica forms and increases the porosity of the clay.

The resulting improvement pattern in the values of CBR in research has good agreement with Meron W and Samuel T. The study concluded that adding 3% lime material increased the CBR value of clay soil from 0.91% to 9.73%. Similarly, Akshaya KS reported that the soaked CBR value increased from 1.6 to 4.1% upon adding 30% ceramic material to clay soil. In the present investigation, however, the 30% WCD showed an improvement from 2% to 8.13%.

Furthermore, the clay soil's CBR Swell was decreased by the addition and increase in stabilizer quantity. The addition of both limestone and waste ceramic dust significantly reduced the CBR Swell of the natural soil. Around 6% natural lime was applied, and the CBR Swell decreased from 6.08 to 4.91. Similarly, applying the same amount of waste ceramic dust resulted in a significant reduction from 6.08 to 4.81. The results of the conducted experimental study also showed that, at dosages less than 6%, the performance of limestone was clearly less significant than that of waste ceramic dust in the same amount. However, the application of limestone continued to be superior to the waste ceramic dust for dosages higher than 6%. In connection with this, the ensuing CBR Swell dropped by 49% and 40%, respectively, for the separate application of 10% natural lime and waste ceramic dust. The slope of the dosage vs. CBR swell graph is comparatively higher for limestone stabilization. It is a useful tool for evaluating how clearly the CBR Swell of natural clay soil is affected by changes in additive dosage. The pozzolanic reaction leads to the development of cementitious materials such as CSH, calcium aluminum silicate, and gyrolite, which decrease clay's sensitivity to moisture and, consequently, its swelling properties. Nifana S. and Palanikumar M. made the observation that there is an inverse relationship between the dosage of ceramic material and the CBR Swell of clayey soils in this regard.

The soil's CBR increased significantly as the dosage of limestone increased, much like the MDD. When the mix combination changes from 15% WCD + 2% limestone to 15% WCD + 6% limestone is increased from 0 to 30%, respectively, the CBR rises from 30 to 20%. Additionally, the CBR Swell decreased as the dosage of stabilizers in each combination increased. As a result, when the amount of natural lime in the combination increased from 2% to 4%—a reduction of 16.1%—the free swell decreased from 4.36% to 3.64%. The strength and frictional resistance of the clay soil increase with an increase in lime dosage.

7.2.3. Effect of Natural Lime and Waste Ceramic Dust on LL and PI of the Clay Soil

When stabilizers are added, the soil's plastic index and liquid limit both decrease in relation to the Atterberg limits. The value of the liquid limit was reduced from 79.6% to 71.45%, 58.27%, and 29.4%, respectively, when 2%, 6%, and 10% limestone were blended with the clay soil. Similar to CBR, the clay soil's sensitivity to natural lime dosage is higher than that of waste ceramic dust, as evidenced by the nearly equal performance of 10% limestone and 30% WCD in lowering the clay's liquid limit. Therefore, about 200% more dosage is needed to achieve a similar degree of improvement in reducing liquid limit with any arbitrary amount of natural lime. This is due to waste ceramic dust.

Similarly, as the amount of both stabilizers increased, the expansive subgrade's plastic index—another significant index property—decreased. The soil's intact plastic index in its natural state was 50.33%. However, the final plastic index value decreased to 14.14 and 7.15%, with the addition of 6% and 10% of limestone, respectively. Likewise, applying the same quantity of waste ceramic dust reduced the stated parameter's magnitude by 3.8% and 10.64%, respectively. The process of flocculation—the close-knit bonding of clay particles resulting from the suppression of pore water—is eventually brought about by the cation exchange that occurs between the additives and the clay material.

According to Subhashree and Bell, flocculation is a crucial process that lowers the clay soil's plastic behavior and, consequently, the liquid limit and plasticity index. As a result, the clay material becomes stronger due to an improvement in its frictional resistance. Rakhil and Devi state that the plasticity index and liquid limit of clay soil decrease as the clay-WCD mixture's ceramic content rises. The study's findings also showed that the liquid limit's decrement rate is significantly slower than the plasticity index's. Additionally, Nifana and Palanikumar discovered that a 30% ceramic material addition to clay soil decreased the liquid limit from 42 to 33%, or a reduction of 21.4%. Research found a similar trend of decline in the liquid limit's magnitude.

When waste ceramic dust and limestone are combined, the amount of the plastic index and liquid limit decreases as the percentage of limestone on the 15% WCD increases. Figure 7.7(a) and (b) show the combined effects of limestone and waste ceramic dust on the clay soil's plastic index and liquid limit, respectively. The natural (stabilized) expansive clay had a liquid limit of 79.6% and a plastic index of 50.32%, respectively. However, for the combination of 15% WCD and 6% natural lime, the post-stabilization values were reduced to 46.83% and 9.88%, respectively. Similarly, after combining the 15% WCD and 4% natural lime with the expanded subgrade, the plastic index value dropped to 10.2%. The combination of 4 and 15% stabilizer dosage was determined to be the most cost-effective way to alter the clay's properties out of all the possible combinations.

The plastic index of the stabilized clay subgrade was significantly lowered by replacing a portion of the plastic clay particles with non-plastic waste ceramic dust and a smaller amount of plastic limestone material. The availability of calcium in limestone and silicate in clay soil, which is crucial for efficient cation exchange, is the primary cause of the decrease in the magnitude of the plasticity index.

Additionally, the ionic exchange between the soil's clay and lime minerals caused the clay particles to aggregate and flocculate, which decreased the soil's plasticity. Meron, Samuel, Rakhil, and Devi noted that when the dosage of lime material increases, the liquid limit and plasticity decrease. Additionally, the characteristics of swelling and shrinkage decrease as the amount of lime material increases.

7.2.4. Effect of Natural Lime and Waste Ceramic Dust on Free Swell of the Clay Soil

stabilizers affected the soil's free swell, just like they did other geotechnical characteristics. The Figure illustrates how increasing and decreasing the quantity of waste ceramic dust and natural lime both led to a reduction in free swell. The expansive properties of the clay soil were considerably changed by the addition of the additives. The addition of water to the clay did not cause an excessive volume change after stabilization of the soil was applied. Consequently, following the independent application of 6% and 10% natural lime, the free swell dropped from 74% to 25.59% and 13.65%, respectively. Similarly, for the same dosage of waste ceramic dust, the value of the stated parameter decreased to 66% and 46.44%. The workability of clay is reduced due to the significant reduction in its plasticity and swelling properties caused by the colloidal reactions that take place immediately after the addition of lime material (Thompson, 1968).

The clay subgrade's expansive behavior was significantly reduced as the amount of limestone material increased. The graph in Figure 7.8 shows that the amount of free swell decreased from 74% to 20% and 10%, respectively, when 4% and 8% of natural lime were added to the waste ceramic dust. Additionally, the combined consideration of additives resulted in a faster rate of decline in free swell than the separate blends. Likewise, the resulting degree of improvement in reducing free swell for the identified optimum dosage of limestone and waste ceramic dust is less than that of the joint action. When applied separately, the 10% limestone caused the free swell to decrease in magnitude by 81.5%, while the combination of 4% limestone and 15% waste ceramic dust reduced the free swell by 88.3%. The research that Kate J.M. et al., showed that adding quicklime (CaO) or hydrated lime ($Ca\ (OH)_2$) to expansive and clayey soils decreases swelling and swelling pressure, which in turn lessens the clay soil's sensitivity to changes in moisture. Similarly, Nifana and Palanikumar found that adding 5% lime reduced the linear shrinkage of clayey soil by an average of 4–7%.

7.2.5. Combined Effect of the Additives on Geotechnical Properties of the Clay

According to the experimental analysis, the ERA 2013 specification for road construction was fully satisfied by the independent application of 6% natural lime in the absence of waste ceramic dust. However, not all parameters exceeded the minimum requirement for the 4% natural lime. Likewise, a minimum dosage of waste ceramic dust of 20% is needed to allow all parameters to fall above the lowest safe margin of the ERA specification. Therefore, it can be concluded that the WCDs of 6% and 20% represent the ideal dosages for applying the materials separately. On the other hand, the combined impact of the two additives was examined to achieve even greater performance improvement from the clay soil. To compare the level of improvement attained in the separate (cases 2 and 3) and combined (case 4) application of the additives, different amounts of natural lime are mixed with 15% waste ceramic dust.

So, in addition to the amounts determined for the independent employment of the stabilizers, the natural lime with dosages ranging from 2% to 10% was blended with 15% WCD to have an alternate optimal dosage. If one of the materials is scarce, the second option might be a good one. Furthermore, it is crucial to apply stabilizers in tandem to take full advantage of the material that performs better. Furthermore, from the perspective of both degree of improvement and having a variety of chemical compositions within the blend, the combined stabilization approach is frequently more effective than the separate consideration of stabilizers. The presence of diverse oxides and chemicals with distinct properties in the mixture promotes the exchange of cations between the stabilizing agents and the clay particles, an essential step in achieving the desired level of improvement.

REFERENCES

1. Abdullahi, S. L., & Audu, A. A. (2017). Comparative analysis on chemical composition of bentonite clays obtained from Ashaka and Tango deposits in Gombe State, Nigeria. *Chemical Search Journal, 8*(2), 35–40.
2. AI-Azzo, S. I. (2009). Treatment of expansive clayey soil in AL-Wahda district at Mosul city with crushed limestone. *INJES, 9*(2), 1–10.
3. Al-Sharif, M. M., & Attom, M. F. (2014). A geo-environmental application of burned wastewater sludge ash in soil stabilization. *Environmental Earth Sciences, 71*(5), 2453–2463.
4. Amadi, A. A., & Okeiyi, A. (2017). Use of quick and hydrated lime in stabilization of lateritic soil: Comparative analysis of laboratory data. *Geo-Engineering, 8*, 3.
5. Amakye, S. Y., et al. (2021). Enhancing the engineering properties of subgrade materials using processed waste: A review. *Geotechnics, 1*, 307–329.
6. Araújo, R. A. (2019). Evaluation of the pozzolanic activity of red ceramic waste using mechanical and physicochemical methods. *Cerâmica, 65*, 461–469.
7. Arif, U. R., et al. (2019). Stabilization of subgrade by using additives (cement, lime). *GSJ, 7*(4).
8. Ayothiraman, R., et al. (2012). Evaluation of liquefaction potential of Guwahati gateway city to North-Eastern India. *Natural Hazards, 63*, 449–460.
9. Bhengu, P. H., & Allopi, D. (2017). Influence of lime-on-lime soil stabilization on natural occurring acidic soil engineering properties. *ASRJETS, 30*(1), 57–69.
10. Binici, H. (2007). Effect of crushed ceramic and basalt pumice as fine aggregates on concrete mortar properties. *Construction and Building Materials, 21*(6), 1191–1197.
11. Damtew, T., et al. (2022). Deformation analysis of cement-modified soft clay soil using finite element method (FEM). *Heliyon, 8*.
12. Edmeades, D. C. (1982). Effects of lime on effective cation exchange capacity and exchangeable cations on a range of New Zealand soils. *New Zealand Journal of Agricultural Research, 25*(1), 27–33.
13. Ellina, B. (2021). Effective cation exchange capacity of calcium silicate hydrates (C-S-H). *Cement and Concrete Research, 143*.
14. Kassahun, A. (2018). Engineered soil and the need for lime-natural pozzolan mixture percentage approximation. *Ethiopian Journal of Science, 41*(2), 70–79.

15. Kumar, B. S., & Preethi, T. V. (2014). Behavior of clayey soil stabilized with rice husk ash & lime. *IJETT, 11*. Retrieved from: http://businessdocbox.com/Green_Solutions/94110833-An-experimental-investigation-on-stabilizing-the-soil-using-rice-husk-ash-with-lime-as-admixture.html (accessed on 5 December 2024).
16. Laurent, de W., et al. (2014). Kinetics of lime/bentonite pozzolanic reactions at 20 and 50°C: Batch tests and modeling. *Cement and Concrete Research, 59*, 34–42.
17. Mateous, M. (1964). Soil lime research at Iowa State University. *Journal of Soil Mechanics and Foundations Engineering, Div. ASCE, 90*(2), 127–153.
18. Mathew, P. K., & Rao, S. N. (1997). Effect of lime on cation exchange capacity of marine clay. *Journal of Geotechnical and Geoenvironmental Engineering, 123*(2), 183–185.
19. Meron, W., & Samuel, T. (2014). Stabilization of expansive soil using bagasse ash & lime. *EEA, 32*. Retrieved from: https://www.ajol.info/index.php/zj/article/download/142172/131911 (accessed on 5 December 2024).
20. Mertens, G. (2009). Pozzolanic reactions of common natural zeolites with lime and parameters affecting their reactivity. *Cement and Concrete Research, 39*, 233–240.
21. Mukesh, A. P., & Patel, H. S. (2012). A review on effects of stabilizing agents for stabilization of weak soil. *IISTE, 2*(6).
22. Mutaz, M., & Muawia, D. (2014). Chemical analysis and X-ray diffraction assessment of stabilized expansive soils. *Bulletin of Engineering Geology and Environment, 73*, 1063–1072.
23. Narasimha, R. S., & Rajesekaran, G. (1996). Reaction products for medium lime-stabilized marine clays. *Journal of Geotechnical Engineering, Geotechnical Div. ASCE, 122*(5), 329–336.
24. Nifana, S., & Palanikumar, M. (2021). Mechanical stabilization of subgrade soil using ceramic powder. *IJME, 6*(3).
25. Oleng, M. (2018). Physical and chemical properties of crushed ceramic and porcelain clay tile powder. *IJIEASR, 7*(7).
26. Prasad, R. R., et al. (2018). Use of lime and waste plastic fibers for subgrade stabilization. *IJEAT, 8*(2C2). Retrieved from: https://www.ijeat.org/wp-content/uploads/papers/v8i2c2/B10091282C218.pdf (accessed on 5 December 2024).
27. Rakhil, K. R., & Devi, K. (2016). Review on the effect of waste ceramic dust on the geotechnical properties of expansive soils. *IRJET, 3*(12).

28. Raymond, N. Y., & Ouhadi, V. R. (2006). Experimental study on instability of bases on natural and lime/cement stabilized clayey soils. *Applied Clay Science, 35*(3, 4), 238–249.

29. Seco, A., et al. (2011). Stabilization of expansive soils for use in construction. *Applied Clay Science, 51*(3), 348–352.

30. Thompson, M. R. (1968). Lime-treated soils for pavement construction. *ASCE, 94*(2), 1991–217.

31. Torgal, J., & Jalali, S. (2010). Reusing ceramic wastes in concrete. *Construction and Building Materials, 24*(5), 832–838.

CHAPTER 8

ENVIRONMENTAL IMPACT AND SUSTAINABILITY IN SOIL STABILIZATION

Contents

INTRODUCTION

In addition to providing structural support, soil stabilization is a crucial building procedure that has several positive environmental effects. We can better appreciate soil stabilization's influence on the environment if we are aware of the concept and how it works in construction. Furthermore, by investigating the link between soil erosion and climate change, we can understand how critical it is to put environmentally friendly solutions—like soil stabilization—into practice. A number of sustainable development objectives, such as life on land (SDG 15), climate action (SDG 13), and responsible consumption and production (SDG 12), are aligned with soil stabilization. It advances the general objective of building a more resilient and sustainable future by halting soil erosion, cutting carbon emissions, and encouraging sustainable land management techniques.

Construction projects can lessen the environmental effect of infrastructure development by implementing soil stabilization techniques. Stabilization techniques can effectively mitigate soil erosion, a major concern on construction sites, while also preserving valuable topsoil and preventing sediment runoff into neighboring water bodies. This helps the nearby communities' access to better-quality water while simultaneously safeguarding the ecosystem. Moreover, soil stabilization is essential to combating climate change. We can lessen the amount of carbon dioxide and other greenhouse gases that are released into the atmosphere by improving soil stability. As carbon sinks, stable soils efficiently store carbon and lessen the effects of climate change. This is especially crucial in regions vulnerable to droughts or other extreme weather, as soil deterioration and erosion can worsen the consequences of climate change.

In short, soil stabilization reduces soil erosion, maintains soil fertility, and lessens the effects of climate change, all of which have a substantial positive impact on the environment. With the use of cutting-edge scientific methods and creative thinking, soil stabilization offers environmentally friendly building and sustainable land management solutions. We can endeavor to protect our environment and build a more sustainable world by realizing its significance.

8.1. ASSESSING THE ENVIRONMENTAL IMPACT OF DIFFERENT STABILIZATION TECHNIQUES

Along with being crucial to modern building practices, soil stabilization has a profound effect on both structural integrity and environmental sustainability. The environmental impact of construction projects is becoming a major concern as global building activity increases. A key ally in reducing adverse environmental effects is soil stabilization, which can be accomplished in many ways, including chemical additions, mechanical techniques, or biological treatments. In addition to strengthening the underpinnings of building projects through the reduction of soil erosion, improvement of water quality, and mitigation of ecosystem disruption, soil stabilization fosters a peaceful coexistence of human development with the delicate balance of the natural environment. To achieve sustainable and ethical building practices, soil stabilization is a crucial tool in this sophisticated building approach that places a strong emphasis on environmental preservation.

8.1.1. Benefits of Soil Stabilization for the Environment

- **Controlling Erosion:** One of the main advantages of soil stabilization for the environment is erosion control. Stabilized soil helps to stop soil erosion, which can lead to the depletion of rich topsoil and cause sedimentation in water bodies. By lowering erosion, soil stability maintains biodiversity, safeguards natural ecosystems, and upholds ecological equilibrium.
- **Improving Water Quality:** Reduced sediment and pollution flow into nearby bodies of water is a result of stabilized soil. The survival of aquatic ecosystems and the preservation of drinking water supplies depend on improved water quality. By limiting sedimentation, soil stabilization helps to preserve the quality of rivers, lakes, and other water resources.
- **Dust Emissions Have Been Reduced:** A significant amount of dust is produced by unstable building sites, which adds to air pollution. Chemical treatments and the use of binders are examples of soil stabilization techniques that can lower dust emissions and create a cleaner and healthier building environment.
- **Construction-Related Disturbance Reduction:** Conventional construction methods have the potential to significantly disturb the surrounding environment, impacting both flora and fauna. By allowing more controlled construction activities, soil stabilization reduces the negative effects on the surrounding environment.
- **Natural Habitat Conservation;** By lessening disturbances to the soil and surrounding ecosystems, soil stabilization contributes to the preservation of natural habitats. This is particularly crucial in environmentally delicate areas where building projects might endanger rare or endangered wildlife.
- **Reduced Carbon Footprint:** Construction projects can lessen their carbon footprint by implementing soil stabilization techniques. By

stabilizing with materials that are readily available locally, emissions related to transportation can be decreased.

- **Resource Efficiency:** Utilizing locally accessible resources for soil stabilization frequently lowers the need for transportation and the energy it requires. This efficient use of resources lessens the impact of material extraction and transportation on the environment, in line with sustainable construction practices.
- **Stability and Durability Over Time:** By providing a solid foundation for building, stabilized soil reduces the possibility of recurring soil-related issues. Because of its long-term stability, building operations have an even smaller negative impact on the environment by requiring less maintenance and repairs.
- **Environmental Regulation Compliance:** Many locations have stringent environmental regulations in place to protect natural resources. These regulations are often followed by soil stabilization techniques, guaranteeing that building projects are completed in an ecologically responsible way.
- **Sustainable Construction Practices Promotion:** Highlighting the ecological advantages of stabilizing soil presents an opportunity to encourage and facilitate the construction industry's adoption of sustainable building practices. In addition to strengthening building foundations, soil stabilization promotes environmental sustainability by reducing disturbances caused by construction, enhancing water quality, and preventing erosion.

8.1.2. Challenges If We Don't Use Soil Stabilization

When soil stabilization is not used during construction, several problems may arise that have an impact on the surrounding environment as well as the building site itself. The following are some of the main problems caused by inadequate soil stabilization.

- **Erosion of Soil:** Without soil stabilization, construction sites are vulnerable to soil erosion, especially in areas with steep slopes or during periods of heavy rainfall. Erosion can result in the loss of topsoil, endangering the stability of the building site and causing deposition downstream in water bodies, which damages aquatic ecosystems.
- **Structural Integrity is Jeopardized:** Unstable soil conditions can compromise the structural integrity of buildings, roads, and other infrastructure. In the absence of appropriate stabilization, soil

settlement, subsidence, and uneven settling may occur, causing long-term damage and raising maintenance costs.

- **Air Pollution and Dust:** When building on untreated soil, a lot of dust is produced, which pollutes the air in the nearby communities. This has greater negative effects on the environment, changing the quality of the air and possibly harming plants, in addition to putting the health of the local populace and employees in jeopardy.
- **Water Pollution and Increased Runoff:** When it rains, untreated soil surfaces are more likely to generate runoff that carries pollutants and silt into nearby bodies of water. This runoff can damage aquatic ecosystems, lakes, and rivers, causing eutrophication that affects human water supplies as well as the environment.
- **Disturbance of Habitats and Biodiversity Loss:** Building on untreated soil can seriously disturb the local flora and fauna by upsetting their habitat. Loss of biodiversity can result from the destruction of natural habitats, nesting sites, and feeding grounds, especially in environmentally sensitive areas. This disturbance has the potential to upset the delicate balance of the environment by contributing to the extinction or displacement of native species.

Lastly, it has become clear that soil stabilization is a crucial component of contemporary architecture, supporting both environmental sustainability and structural reinforcement. The benefits of soil stabilization, which include reduced carbon emissions and erosion control, emphasize how important it is to encourage ethical building practices. Refusing to accept soil stabilization poses serious problems, endangering structural integrity, causing pollution of the air and water, and upsetting ecosystems. Soil stabilization firms employ soil stabilization practices as a crucial step toward achieving sustainable development and fostering a harmonious coexistence between human progress and the delicate balance of the natural world. This is particularly relevant as the construction industry struggles with the imperative of balancing growth and environmental preservation.

8.1.3. Environmental Impact of Different Stabilization Techniques

8.1.3.1. Chemical Soil Stabilization

Chemical stabilization is the process of improving soil properties by adding chemical agents. These substances, which include fly ash, cement, and lime, interact with the soil particles to increase stability and cohesiveness. Cohesive soils like clay are frequently stabilized chemically to improve their strength, decrease their plasticity, and reduce their absorption of water. The benefits of chemical stabilization include its ability to treat quickly and its *in-situ* application, which minimizes the need for significant soil replacement and excavation. The procedure is also environmentally friendly because chemical agents like lime and fly ash are frequently leftovers from other industrial operations. Moreover, the type and concentration of chemical agents used in chemical soil stabilization can be changed to suit particular soil conditions. With this customization, engineers can solve a variety of soil issues, such as enhancing load-bearing capacity and lowering settlement over time.

8.1.3.2. Mechanical Soil Stabilization

The process of mechanically stabilizing soil entails physically modifying its characteristics. Compaction, soil confinement systems, and geosynthetics are some of the techniques that strengthen and stabilize the soil. Compaction techniques are used to increase the density and load-bearing capacity of soil layers by compacting them. Systems for confining soil, like geocells or geogrids, form a structure resembling a grid that strengthens and confines the soil, stopping erosion and lateral movement. Similar to geotextiles, geosynthetics are used to lessen water-related problems by enhancing filtration and drainage capabilities. Furthermore, to get the best results, mechanical soil stabilization techniques are frequently combined with chemical ones. Through the integration of both methodologies, engineers can effectively utilize the advantages of each approach to formulate a comprehensive strategy for soil stabilization that concurrently tackles several soil challenges.

8.2. SUSTAINABLE PRACTICES IN SOIL STABILIZATION

It is imperative that all industries contribute to the adoption of sustainable practices in a world where environmental consciousness is growing. Road construction is one industry that is frequently left out of discussions about sustainability. Because they act as the arteries connecting communities and promoting economic growth, roads are an essential component of modern society. However, the environmental cost of using traditional road construction methods can be significant due to resource consumption and carbon emissions. Soil stabilization enters the picture in this situation. In this section, we will explore sustainable road construction solutions in more detail and advocate for the revolutionary use of soil stabilization as a way to build affordable, environmentally friendly road infrastructure.

8.2.1. The Dilemma of Road Construction

Conventional methods of building roads require a significant amount of non-renewable resources, like concrete, asphalt, and gravel. These actions contribute to climate change because they use a lot of resources and produce a lot of carbon emissions. The ongoing need for upkeep and repairs also makes the environmental impact worse. These practices have a significant negative impact on the environment, so it's time to look into more sustainable ways to offset them.

- **Depletion of Non-Renewable Resources:** A significant amount of non-renewable resources is used in traditional road construction. Since gravel is essential to road foundations, large-scale quarrying operations are frequently used to obtain it, resulting in the loss of natural ecosystems and landscapes. Furthermore, the extraction of raw materials is required for the production of asphalt and concrete, which not only uses a significant amount of energy but also disturbs ecosystems.
- **Climate Change and Carbon Emissions:** One major worry is the carbon emissions brought on by conventional road construction. Significant greenhouse gas (GHG) emissions are produced during the whole life cycle of these roads, from material extraction through transportation, processing, and construction. These emissions' rapid atmospheric release causes climate change, which in turn causes unfavorable weather patterns, an increase in global temperatures, and other disastrous environmental effects.
- **Ongoing Maintenance and Repairs:** Traditional road construction has an ongoing negative environmental impact even after the initial stages of construction are completed. Regular upkeep and repairs are necessary for roads constructed with materials that are heavy on resources, such as concrete and asphalt. The ongoing reconstruction cycle intensifies the environmental impact even more. It entails using more resources, producing more emissions, and upsetting the ecosystems nearby.

8.2.2. Sustainable Road Construction with Soil Stabilization

The goal of sustainable road construction is to reduce these adverse effects while maintaining the efficiency, safety, and durability of the road infrastructure. It is important to consider a road's whole life cycle, from construction to maintenance to eventual decommissioning, rather than just the finished product. Soil stabilization is a novel and efficient way to accomplish this aim.

8.2.2.1. Natural Fiber Reinforcements in Soil Stabilization

Because natural fibers are renewable and biodegradable, they are essential to sustainable soil stabilization techniques. In terms of engineering properties, these reinforcements provide a sustainable and eco-friendly way to improve soil for building and infrastructure projects. By using them in soil stabilization, the environmental impact of traditional stabilization techniques is lessened and an efficient and sustainable way to improve soil strength, cohesion, and stability is provided. Reinforcements made of natural fibers that come from replenishable natural resources are both reasonably priced and environmentally beneficial when compared to synthetic materials. Their use encourages improved soil health and erosion control while cutting costs, which is consistent with the principles of sustainable construction. A more environmentally friendly approach to soil stabilization can be achieved by incorporating natural fiber reinforcements into the process, which will enhance the overall sustainability of infrastructure and construction projects. To fully utilize natural fiber reinforcements in soil stabilization applications, their engineering properties must be understood and applied effectively.

8.2.2.2. Advancements in Eco-Friendly Geosynthetics

An important development in sustainable soil stabilization techniques is the use of environmentally friendly geosynthetics, which provide a workable substitute for conventional techniques and are consistent with the ideals of ecologically responsible infrastructure development and building. The development of environmentally friendly geosynthetics has several advantages for engineering uses, environmental effects, and soil stabilization applications. A few crucial elements are as follows:

- Biodegradable polymers lessen the accumulation of non-recyclable waste and promote sustainability.
- The performance of geosynthetics is improved by developments in bio-based reinforcement fibers, regulated degradation processes, and engineered biodegradability.
- The growing need for environmentally friendly substitutes highlights how crucial it is to develop sustainable infrastructure.
- In geosynthetics, the use of waste materials like rice husk ash and recycled waste promotes a circular economy and lessens environmental impact.
- In stabilized soil applications, the mechanical properties of eco-friendly geosynthetics offer enhanced soil health and cost-effectiveness due to their flexibility.

8.2.2.3. Utilizing Waste Materials for Stabilization

The use of waste materials for soil stabilization offers a practical and affordable way to lessen the environmental impact of building projects. Fly ash, rice husk ash, and palm oil fly ash are examples of waste materials that are being used more and more in soil stabilization projects as a practical substitute for more conventional techniques like making cement and lime. When utilized to stabilize soils, these materials exhibit promising engineering qualities, such as increased compressive strength and load-bearing capacity. The construction sector can greatly lessen its environmental impact and support sustainable development by utilizing waste materials in soil stabilization. These techniques reduce the need for new resources while also helping to recycle waste, which decreases the harm that comes to the environment. The potential of waste materials for stabilizing soil is highlighted in the Table 8.1, highlighting their benefits for infrastructure projects and the broader sustainability movement.

Table 8.1. Risks, Possibilities and Advantages of the Recycling and Reuse of Building Material Types

Type of Material	Risks Related to Building Material Remainders	Possibilities and Advantages Related to the Reuse of Recycled Material
Wood	• Lingering in landfills, which may cause harmful putrefactive processes when stored improperly. • Loss of valuable biological material.	• Glued wood from recycled waste as a component of high-strength structures. • Creation of new mixed materials with higher strength properties.
Plastic	• Generation of additional waste, occupying a large area of land. • Risk of additional soil and air contamination during long-term storage.	• After recycling—as a component of concrete and rubber used in construction. • In production of "green bricks."

Concrete	• Waste that uses large amount of space. • With improper storage, there is a risk of additional soil contamination.	• With proper disassembly—the possibility of reusing the elements. • After recycling—a component of concrete and asphalt mixes.
Glass	• Waste that poses a potential risk of physical damage during the organization of people's work in storage and deposition. • Increased fire hazard during storage (lens effect).	• Recycled glass can be a component of cement and concrete mixes. • Special properties of "glass powder" (GP) when connecting concrete with steel construction elements.
Ceramics	• A very large storage area for this type of construction waste. • Waste that poses a potential risk of physical damage during the organization of people's work in storage and deposition.	• Reuse as fine aggregate and as a substitute for cement in concrete mixes. • Covering material on façades and internal walls. • Flooring material inside buildings and in urban spaces.
Metal	• Long-term storage poses a risk to the environment due to chemical decomposition, which can be a threat to soil and climate.	• Due to the specificity of the material, a large percentage can be transformed into reused building elements. • Self-compacting concrete ingredient.

Source: Rybak-Niedziółka K, Starzyk A, Łacek P, Mazur Ł, Myszka I, Stefańska A, Kurcjusz M, Nowysz A, Langie K. (2023). Use of waste building materials in architecture and urban planning—A review of selected examples. Sustainability, 15(6):5047. https://doi.org/10.3390/su15065047.

8.2.2.4. Sustainable Solutions for Soil Improvement

Sustainable soil improvement techniques are essential for boosting soil health while maintaining environmental sustainability. What sustainable options are available for enhancing soil stability and strength in construction projects? The engineering qualities of soil are enhanced by many cutting-edge methods and stabilizing agents, which makes them perfect for infrastructure projects like building new roads. Among the best long-term options for improving soil are the following:

- **Natural Fibers:** Increasing soil strength, cohesion, and stability, natural fibers are a cost- and resource-efficient eco-friendly substitute for synthetic stabilizers.
- **Bio-Based Geosynthetics:** With regulated breakdown processes, bio-based geosynthetics present a bright future for sustainable soil stabilization that lessens long-term costs and environmental impact.
- **Enhancement of Load-Bearing Capacity:** Sustainable soil stabilization is a perfect and economical option for building projects because it offers advantages like decreased erosion, better water quality, and increased load-bearing capacity.
- **Enhancement of Soil Shear Strength:** By enhancing the soil's shear strength, these techniques also help to guarantee the stability and longevity of the built infrastructure.
- **Erosion Control:** By reducing erosion, effective soil stabilization techniques protect the environment and guarantee long-term sustainability.

8.2.2.5. Environmental Benefits of Soil Stabilization

Reducing soil erosion, maintaining land integrity, and safeguarding water quality are all made possible by soil stabilization techniques, which also promote environmental sustainability. Sustainable techniques for stabilizing soil improve its engineering qualities, which benefits the environment as a whole. Because stabilized soil can support more weight, it resists erosion better. This keeps the land intact for building and agriculture, as well as preventing the loss of priceless topsoil. Stabilized soil also protects aquatic ecosystems and water bodies by retaining pollutants and sediments. Additionally, utilizing eco-friendly products for soil stabilization lowers carbon emissions related to transportation and the creation of new resources. Soil stabilization offers a sustainable and economical solution by reducing the need for costly excavation and the replacement of inappropriate soil. In the end, these procedures lessen the negative effects that building operations have on the environment and support long-term environmental sustainability. Thus, the advantages of soil stabilization for the environment go beyond mitigating erosion; they also include maintaining soil structure, enhancing soil quality, and safeguarding water quality.

8.2.2.6. Innovative Approaches for Eco-Friendly Soil Stabilization

The advantages of soil stabilization for the environment, especially when it comes to fostering sustainability and maintaining soil integrity, provide the groundwork for investigating novel strategies for environmentally friendly soil stabilization. There are now many cutting-edge methods for stabilizing soil that are also environmentally friendly and have many benefits for the building sector. These methods consist of:

- Strengthening and cohesiveness of soil by using natural fiber reinforcements, which are biodegradable and renewable.
- Combining biodegradable polymers and other environmentally friendly geosynthetics to enhance soil stabilization's mechanical properties' flexibility and sustainability.
- Analyzing degradation rates and the suitability of natural fiber reinforcements for various soil types to guarantee reliable integration methods and constant quality.
- Tackling issues related to sustainability to fully utilize eco-friendly geosynthetics and natural fiber reinforcements in soil stabilization projects.
- Encouraging healthy soil and providing sustainable soil stabilization techniques that save money on environmental cleanup and upkeep over the long run.

Novel strategies for environmentally friendly soil stabilization emphasized cost-effectiveness and sustainability in building projects, all the while enhancing the soil's engineering qualities. Fly ash, a coal combustion by-product, is another sustainable method being investigated to enhance soil stability.

8.2.3. Towards Sustainable Soil Stabilization in Peatlands: Secondary Raw Materials as an Alternative

The building sector needs to change and adopt sustainable practices as the globe struggles with environmental issues. A creative answer to the problem of building roads is soil stabilization, which enables the creation of long-lasting and ecologically friendly roads. It's a strategy that balances environmental responsibility with economic considerations following the principles of sustainability. Soil stabilization is a feasible alternative that governments, construction companies, and individuals should take into consideration to make sustainable road construction a reality. This approach can help us create a more economical, ecologically conscious, and sustainable future. We can all work together to bring about positive change and make the world a more sustainable place for the coming generations. One stabilized soil road at a time, let's take the less-traveled route and construct a better, more sustainable world. The road under your feet is where sustainability starts, and we can create roads that benefit us now and protect the environment for future generations by using creative solutions like soil stabilization. Sustainable road construction can have a big impact on how we develop and maintain our vital infrastructure because it is not just an idea but also a workable and realistic reality. It's time for the building sector to adopt these environmentally friendly methods and open the door to a more promising future.

8.2.3.1. An Overview of Peatlands as Secondary Raw Materials

Over 4 million km^2 of the planet is covered by peatlands, which are primarily responsible for providing ecological services and functions that are important to the entire world.

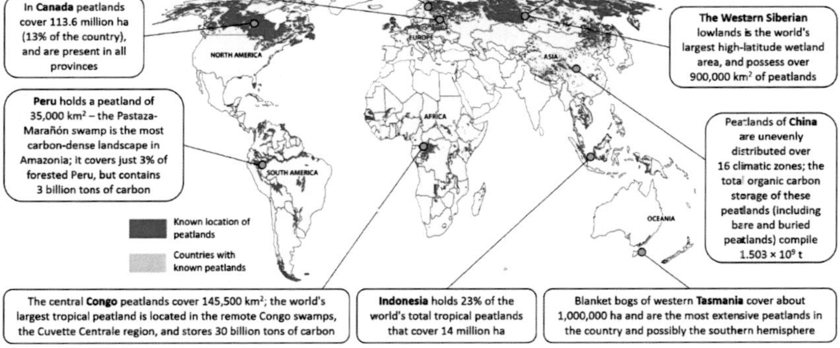

Figure 8.1. *Distribution of peatlands worldwide and complementary facts.*

Source: Vincevica-Gaile, Z.; Teppand, T.; Kriipsalu, M.; Krievans, M.; Jani, Y.; Klavins, M.; Hendroko Setyobudi, R.; Grinfelde, I.; Rudovica, V.; Tamm, T.; et al. (2021). Towards sustainable soil stabilization in peatlands: Secondary raw materials as an alternative. Sustainability, 13, 6726. https://doi.org/10.3390/su13126726.

These services and functions include flood control, water storage and filtration, coal (peat) management, carbon storage and sequestration, transpiration cooling, wildlife habitat, and recreational areas (Joosten, 2016). Bogs, swamps, marshes, mires, and wetlands are typically involved. Peatlands extend from subarctic to equatorial zones, although the largest areas are concentrated in the northern regions (Figure 8.1). However, due to regional variations in classification details and definitions, not all peatlands may be accurately identified. Peaty soils are Histosols, according to the Food and Agriculture Organization of the United Nations (FAO).

Due to their unique geological and environmental characteristics—weak and unstable soil basis, high moisture content, sparse vegetation, unsuitable environmental conditions for agriculture and settlements, and restricted accessibility—peatlands were for centuries considered worthless, low-value areas (Johansen et al., 2017). Nonetheless, peatlands became more urbanized as a result of industrial development. For instance, peatlands are used to build the airports in Zurich and Kuala Lumpur, as well as cities like Saint Petersburg and Amsterdam. Peatlands are becoming more and more exposed to industrial activities due to the expanding global population and the need for land. Large tracts of peatlands are found next to rivers and on the coast, which emphasizes the need for infrastructure to build towns, ports, and resorts. This infrastructure includes roads, highways, trains, airports, power lines, and other access points. Peatlands are used on an industrial basis to extract fossil resources, primarily oil, gas, and peat, which cannot be done without the right infrastructure for maintenance, transportation, and electricity. Undoubtedly, industrial operations negatively impact the ecology and environmental preservation of peatlands; however, if they are unavoidable, they can be carried out more sustainably than in previous decades.

High temperatures and humidity, coupled with heavy precipitation and soil erosion, have an impact on peatlands in equatorial and tropical regions. On the other hand, people living in subarctic areas are subject to unique effects from seasonal fluctuations in frost, permafrost that occurs naturally, and permafrost deformation brought on by climate change. Due to the unique climatic and geological conditions of each site, traditional soil stabilization techniques and technologies must be updated and modified, sometimes involving the discovery of new materials. Better, more affordable, and long-lasting infrastructure construction is required for the world's economy to grow, mostly because mining and forestry are essential. Thus, creative, sustainable methods of stabilizing soil that incorporate the concepts of the circular economy are required. The circular economy, in its ideal state, calls for all waste that is produced, whether it comes from a city, country, or industrial source, to be recycled and used as a secondary raw material. These materials are generally managed similarly to primary raw materials, i.e., the European Union (EU) has the same laws governing trade and transportation; nevertheless, to encourage material sorting, recycling, and reuse, waste-related laws need to be changed. It is challenging to use secondary raw materials widely, even though quality standards are still being developed. Nevertheless, it is only a matter of time, as the scarcity of resources is putting pressure on society and industry stakeholders' wallets. Valuing secondary waste materials is in line with the UN's Sustainable Development Goals and the Circular Economy Action Plan. By creating novel composite materials and examining how secondary materials, like waste or ashes from industry, might be valued and turned into value-added products for stabilizing soils, the life cycle of primary source materials is extended, closing the circle on material recycling.

8.2.3.2. Challenges in Peatland Stabilization

8.2.3.2.1. Demand for Stabilization

Engineers face challenges when building infrastructure outside of zones with stable soils and good bearing capacity, and there is a noticeable search for new materials and technologies. If preloading, floating roads on geogrid reinforcement, or piled embankments cannot be used on weak soils, soil stabilization is required; otherwise, safe engineering planning is nearly impossible, making the construction of infrastructure that is absolutely necessary. Several reasons emphasize the importance of peatland stabilization, i.e., peaty soils generally inhere harsh geotechnical conditions due to the low value of undrained shear strength at natural conditions, high water content, and low permeability, as well as low strength in combination with high compressibility and shrinkage on drying.

In terms of compressibility, "uncontrolled or unexpected decomposition in fibrous peat deposits that may cause significant additional settlement of bearing strata, adversely impacting on the performance of engineering structures founded on or within such deposits" is what defines peaty or boggy areas. The primary factor influencing the mechanical properties of peat is its organic matter content, according to research on decomposition rates. Compression can make peat stronger and less compressible, but it can also cause water loss when the irreversible process of drying out leaves the peat with little moisture absorbed in comparison to its original amount.

Implementing soil stabilization entails modifying and improving the physical characteristics of the soil for engineering objectives; nevertheless, it may also have an impact on the chemical composition of the soil and nearby waters (O'Kelly and Pichan, 2013).

8.2.3.2.2. The Problem of Permafrost

Peatlands in subarctic areas—the north of Europe and Asia, Alaska, Canada, and the Arctic Ocean islands—are subject to the unique effects of permafrost; additionally, in the circumpolar permafrost region, peatlands make up nearly 20% of the land area. The permafrost is usually found beneath the active layer in permafrost regions, with the uppermost layer—known as the active zone—frequently experiencing freeze-thaw cycles. Soil temperature typically stays below freezing for several years under permafrost conditions, but if environmental changes bring about a warmer climate, the soil's mechanical and physical strength will irreversibly decline.

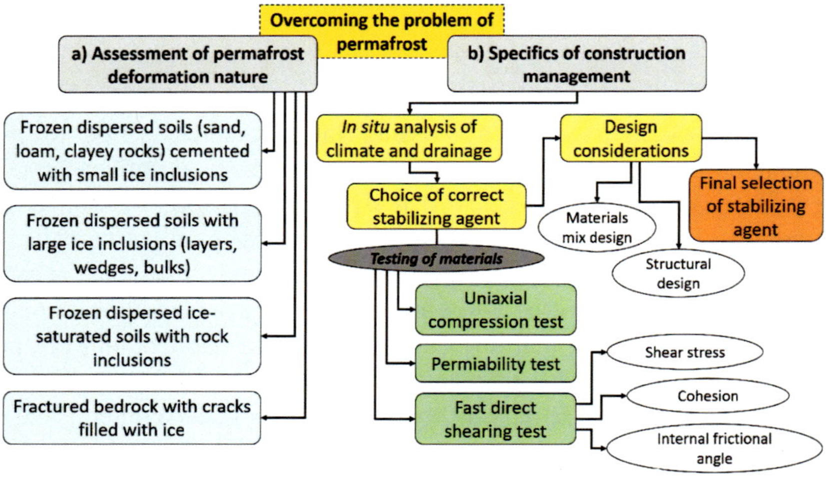

Figure 8.2. *Permafrost deformation nature (a); and its construction management specifics at stabilization (b).*

Source: Vincevica-Gaile, Z.; Teppand, T.; Kriipsalu, M.; Krievans, M.; Jani, Y.; Klavins, M.; Hendroko Setyobudi, R.; Grinfelde, I.; Rudovica, V.; Tamm, T.; et al. (2021). Towards sustainable soil stabilization in peatlands: Secondary raw materials as an alternative. Sustainability, 13, 6726. https://doi.org/10.3390/su13126726.

Over time, the equilibrium state has been reached by ice and voids, fine soil, and mineral and organic matter in undisturbed permafrost. However, this equilibrium shifts and permafrost deformation happens as a result of any engineering construction of infrastructure (Figure 8.2a). For instance, clearing vegetation causes permafrost to thaw more quickly, and building embankments with a lower albedo allows more heat to reach the permafrost beneath. If mass stabilization is intended, these are the permafrost challenges that call for specialized construction management (Figure 8.2b). Numerous studies highlight how permafrost degradation is accelerated by rising ambient temperature, how vulnerable these regions are, and how there is no sign that this situation will change.

8.2.3.2.3. Mass Stabilization as an Option

In Finland and North America, mass stabilization technology—also referred to as a "Nordic approach"—has been widely applied for decades. It can be used in specific situations, like permafrost areas, which are common in peatlands in particular latitudes, as well as for routine soil stabilization. With the use of this technology, the soil bearing capacity beneath roads, railroads, and warehouses is increased. Furthermore, if pollutants persist in formerly industrialized or other heavily anthropogenically influenced areas, mass stabilization is also functional for their mechanical and chemical binding (Scope et al., 2021).

The guidelines state that the Nordic approach to peat stabilization uses a cement mixture that includes gypsum and/or furnace slag. Cement and oil shale ash have been shown to be successful soil mixture binders in the Baltic region. In addition to the binder materials, it is advisable to mix the treated peat layer with different subcomponents like sand or clay because fibrous peat yields the stabilization process more readily than amorphous peat. Prehomogenization, the initial stabilization stage, entails incorporating the constituents into the soil to enhance the ultimate outcome of mass stabilization. Additional challenges arise with stabilization, where a more thorough analysis is needed based on detailed studies of drainage, climate, and design.

Mass stabilization has been used for more than 25 years in more than 30 countries as a viable technique for improving weak soils, including areas covered by permafrost. As a result, mass stabilization has a highly developed technical foundation. Since the equipment is portable and reasonably light, it can be used in a variety of settings and is easily transportable. After conducting a thorough geological, climatic, and other assessment, a specific mass stabilization strategy (Figure 8.3) at a specific location must be determined. For instance, mass stabilization in the shape of triangle piles uses less material than other options, such as stabilization in the form of blocks or columns (Vincevica-Gaile et al., 2021).

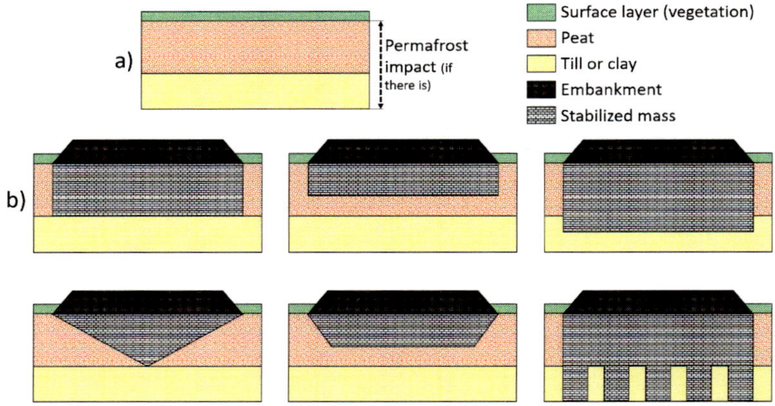

Figure 8.3. *Schematic cross-sections of approach for mass stabilization in peatlands—nonstabilized peatland (a); and variations of stabilization (b).*

Source: Vincevica-Gaile, Z.; Teppand, T.; Kriipsalu, M.; Krievans, M.; Jani, Y.; Klavins, M.; Hendroko Setyobudi, R.; Grinfelde, I.; Rudovica, V.; Tamm, T.; et al. (2021). Towards sustainable soil stabilization in peatlands: Secondary raw materials as an alternative. Sustainability, 13, 6726. https://doi.org/10.3390/su13126726.

As far as geotechnical concerns go, the hydration process that occurs after the binder (e.g., slag or fly ash) is combined with the soil. Extended investigations demonstrate that stabilized soils gained 1.6 times their original strength, and the procedure is economical in regions with limited transportation options for removed peat and limited supply of building materials for soil replacement. According to a study, the cost of a binder can affect mass stabilization unit prices by as much as 70%. Conventionally used cement is costly and has a significant carbon footprint. The substitution of binders made from secondary raw materials, like industrial waste (e.g., gypsum, fly ash, oil shale ash, furnace slag powder, etc.), was put to the test in both the field and a lab. The mass stabilization of peatlands with fly ash-based binders was supported by the leaching test results.

8.2.3.3. Applying a Circular Economy Approach in Material Choice for Stabilization

A growing number of people are focusing on more sustainable businesses and advocating for new regulations that incorporate sustainability principles, particularly in material recycling, due to growing environmental concerns, government support, and mindset shifts. Reducing waste and creating long-term value for materials are the main goals, and designing for (i) recycling; (ii) remanufacturing and reuse; (iii) disassembly and sorting; and (iv) the environment is crucial. More rapid production changes toward the compromise between environmental and economic goals are being driven by stronger regulations on energy efficiency and pollution. The need for virgin raw materials, which are typically obtained from ore mining, may be greatly reduced by substituting secondary raw materials for primary raw materials. Application of different waste materials: natural fibers, alum waste, fly ash from the energy industry, fuel ash from palm oil, etc., for soil stabilization, have been investigated and documented in scholarly works. Geotechnical engineering solutions must be both economical and environmentally friendly to comply with the circular economy approach. The analysis of the industrial loop and the availability of secondary raw materials from nearby industries should, however, inform the decision made on the spot. In addition to conventional materials, an overview of studies conducted in the last few decades offers information on a wide range of secondary raw materials that can be used for soil stabilization. The strategic use of secondary raw materials to cut waste production is something that European governments and multilateral organizations are becoming more and more interested in.

It is clear that, depending on the economic activity of the area, different constituents are used for soil stabilization as a binder or amendment material. Nonetheless, the utilization of secondary raw materials ought to be integrated into the regular administration of infrastructure advancement. Large volumes of some possible soil stabilization amendments are produced by industries. Since the goal is to lessen the hazardous effects of waste masses while also reducing the use of primary raw materials (cement, lime, and sand), which require additional mining and production activities, the decision to choose additional material recycling over waste disposal is crucial.

Because conventional methods for stabilizing peatlands have been developed and proven, they are always applicable. However, beneficial management suggests minimizing the use of primary resources and lowering GHG emissions while building. Research has prompted experiments with a comparatively broad range of raw materials obtained from mining and local resource extraction. However, because they are widely available and inexpensive, using secondary raw materials—a variety of waste materials and byproducts—is a promising option. One method is stabilizing the soil using waste materials, such as shredded specific solid waste or portions of municipal waste, and converting them into useful materials for the construction of infrastructure. Additional materials include geopolymeric binders, which are used as building materials instead of Portland cement. Combustion ashes, metallurgical slag, and glass processing waste are a few examples of solid waste types that could be used to make building materials with less energy and carbon footprint. Alkali-activated materials are primarily low-calcium blast furnace slag and coal fly ash varieties. These materials have qualities like high compressive strength, low heat conductivity, acid and/or fire resistance, etc., that are required for particular applications. Although fly ashes of class C with high calcium content have also been studied, fly ashes of class F with low calcium content are most frequently used to synthesize geopolymeric materials.

Alkali activation's ability to valorize a variety of solid industrial wastes is a promising technique for repurposing specific disposed residuals. The primary aluminosilicate structure of the material is altered by the addition of an alkali source (activator), resulting in secondary polymeric silicate binder phases that enhance the cementitious qualities of the subject material. Based on alkali-activated materials, mortar and concrete have the mechanical performance of regular Portland cement systems, high fire and acid resistance, and the ability to solidify and stabilize other wastes, including heavy metal immobilization (Vincevica-Gaile et al., 2021).

8.2.3.4. Peculiarities Regarding the Use of Certain Secondary Raw Materials

8.2.3.4.1. Solving the Problem of Oil Shale Waste

Low-quality fossil fuels, such as oil shale, are present all over the world, but only a few nations (e.g., Brazil, China, Estonia, Jordan, and Israel) use them on an industrial basis. The primary cause is associated with the low calorific values of these fuels, in addition to technological and ecological problems. The oil shale industry has produced a lot of waste over the years, mostly oil shale ashes. This has led to large white mountains in northern Estonia, which may have hazardous leaching effects on the surrounding environment. Since kukersite oil shale has been mined for almost a century, Estonia is the world's largest producer of solid oil shale. This fuel has relatively low calorific values, and during production, 40–50 weight percent of solid waste is produced (Figure 8.4), of which 98% are landfilled. Furthermore, processing leaves behind 45–50% solid waste (oil shale ashes), which has little use in the secondary market. The use of oil shale as a fuel in power plants comes with many issues, one of which is the massive amount of alkaline ashes that are left over after burning oil shale and are deposited nearby. Estonian thermal power plants produce over 7 million tons of ash annually.

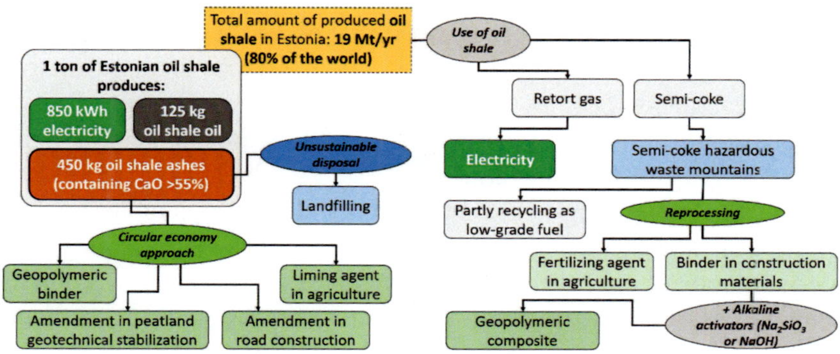

Figure 8.4. *Processing of oil shale, indicating the applicability of the circular economy approach.*

Source: Vincevica-Gaile, Z.; Teppand, T.; Kriipsalu, M.; Krievans, M.; Jani, Y.; Klavins, M.; Hendroko Setyobudi, R.; Grinfelde, I.; Rudovica, V.; Tamm, T.; et al. (2021). Towards sustainable soil stabilization in peatlands: Secondary raw materials as an alternative. Sustainability, 13, 6726. https://doi.org/10.3390/su13126726.

Oil shale has varying amounts of organic, carbonate, and terrigenous material, but its high calcium content—up to 55%—allows for the useful use of this waste. Its consistency is comparable to fly ashes produced by burning coal, and it can be used for both agricultural and road construction. Around 95% of the material is dumped in sizable ash fields close to power plants. One fraction from the oil shale processing can be added to regular Portland cement; this fraction has higher concentrations of SiO_2 and Al_2O_3 than other solid waste. Other ash fractions, on the other hand, are subjected to hydraulic transportation in sedimentation ponds in a slurry with a water-to-ash ratio of 20:1, resulting in the creation of mountain plateaus in Estonia that cover an area greater than 20 km^2 and hold more than 300 Mt of ash. Although there has been a lot of research done on the physical, chemical, and mineralogical characteristics of oil shale ashes, there is still little secondary use. Through the use of analytical techniques, the geopolymeric potential of oil shale ashes for alkaline activation under ambient conditions and construction applications has been investigated. This has resulted in the formation of an amorphous C-(A)-S-H gel, which gives the samples eight times higher compressive strength (8.25 MPa) than samples that exhibit cementitious properties upon hydration.

The concentration of trace elements in oil shale ashes is deemed acceptable for industrial areas, except for arsenic, which is detectable in ashes obtained from electrostatic precipitators. This type of ash contains approximately 40 mg/kg of arsenic, compared to the permissible limit of 20–30 mg/kg for residential areas and 50 mg/kg for industrial zones. Only 23% of the total amount of ash produced, nevertheless, comes from precipitators. The main uses of oil shale ashes are still in soil liming in agriculture, as well as amendments as Portland cement additives and filter material for wastewater treatment. However, the waste is unsuitable for long-term construction and/or stabilization applications due to the weak and chemically unstable cementing bonds that are formed upon simple hydration of Ca-rich oil shale in aggressive environments and continuous leaching. Global usage of single oil shale is expected to increase, particularly for oil and gas production. To develop new, environmentally friendly, and sustainable uses for this waste, it is imperative to enhance the cementitious properties of the waste. Alternatively, fly ashes from oil shale have the right qualities to be mixed with other organic materials like peat. If the only raw materials used in advanced construction materials are waste and local materials, then they are environmentally sustainable.

A partially renewable resource, peat continues to build up on 60% of the world's peatlands; global peat reserves are estimated to be 12 billion tons, spread across an estimated 400 million hectares of peatlands; in 2014, only about 30 million tons of peat were extracted. Several locations across the world sell oil shale ashes or similar ashes from coal or lignite power plants. Peat organic matter and oil shale ashes combined offer superior compressive strength (approximately 1.2 MPa), bending stress, and thermal resistance, making them suitable for use as building materials. However, if oil shale ashes are used in road construction, it is advised to keep an eye on the concentration of trace elements and leaching parameters near surface waters. Surface water discharges, however, necessitate extensive modeling and computations.

The concentration of sulfates and barium in the nearby surface water has been impacted by road construction, according to a case study on the use of oil shale ashes for road construction in an eastern Estonian swamp. Due to the significant geotechnical challenges, new pilot studies are being conducted in wetlands to test different structural design schemes and determine whether oil shale ashes, as a stabilization material, could be used in smaller quantities by altering the design of stabilized masses.

Using the material reduction scheme might be even more stable, even if it doesn't worsen the bearing stability parameters. This approach would also lead to qualitatively less leaching because less material amendment would be required. These tests are being conducted in Estonian peatlands for the construction of new forestry roads using fly ashes from oil shale, peat, and water only—cement is not added.

Another hazardous waste that comes from the oil shale retorting process is oil shale semicoke. The world's petroleum industry is currently experiencing a historic shift from conventional to unconventional gas and oil, and the waste product known as oil shale semicoke poses serious environmental risks. The problem of long-term accumulated oil shale semicoke in landfill sites could be greatly alleviated by using semicoke as a road base or subbase course construction material. Semicoke can be partially disposed of by recycling it to some extent and burning it as a low-grade fuel. Semicoke can also be reprocessed and used as a raw material for building materials and fertilizers. By enhancing the pozzolanic reaction with cement, the active ingredient in semicoke may enable the material to be used as an aggregate in cement-stabilized road base or subbase course construction, thereby reducing the need for natural gravel resources and transportation expenses. Tests such as the raw material test, modified compaction test, splitting tensile strength test, compressive resilient modulus test, unconfined compressive strength (UCS) test, and freezing-thawing test have demonstrated that oil shale semicoke is suitable for use as an amendment for base course material of a low-grade highway or subbase course material of a highway.

8.2.3.4.2. Applicability of Municipal Solid Waste Bottom Ashes in Weak Soil Stabilization

When municipal solid waste (MSW) is incinerated, a substantial amount of combustion residuals are produced, with bottom ashes accounting for roughly 20% of the initial mass. Over 30 million tons of MSW (or 13% of all domestic MSW) were burned in the USA alone in 2014; this equates to 5 million tons of residual bottom ash. Fly ashes are mostly mixed with bottom ashes to prevent them from being labeled as hazardous waste. Bottom ashes, in contrast to fly ashes, are still primarily disposed of rather than utilized as a secondary raw material. A great deal of research has been done to find useful uses for MSW bottom ashes, mainly as a secondary raw material in infrastructure. Bottom ash recycling has been practiced in Europe and Asia and is gradually gaining traction in the United States of America for use as an aggregate material in road construction.

Reusing bottom ash on roadways is not only a common practice but also frequently required by law in several EU nations. If traditional aggregates can be supplemented or replaced in roadways, MSW bottom ashes may prove economically advantageous (especially when accounting for the avoided cost of landfilling). MSW bottom ashes exhibit similar mineralogy and physical properties to some natural aggregates. Moreover, the incorporation of recycled aggregate sources into road bases reduces reliance on naturally extracted minerals and mitigates GHG emissions linked to the extraction of virgin minerals. The Toxicity Characteristic Leaching Procedure primarily classifies MSW bottom ash as a nonhazardous waste in the United States. Additionally, research has shown that MSW bottom ash includes trace amounts of metals and salts that could be harmful to the ecosystem and the environment. If this material is positioned beneath a layer of compacted pavement, the risk of direct human contact with it is minimal; nevertheless, one must still take into account the possibility of contaminants seeping into groundwater.

There is potential for the future use of bottom ashes from incinerators combined with regular granular materials for road base (without diluting hazardous waste, of course), and the legal implications need to be taken into account. However, there's a chance that the amount of trace elements that leach isn't always equal to the initial mass of ash present. The findings provide credence to the idea that mixing MSW bottom ashes with natural and recycled aggregates could be a useful engineering control to reduce the amount of potentially hazardous and toxic elements that are leached into the environment. Because aggregate requirements in road construction are higher than ash quantities, blending could open up previously unfeasible options for ash recycling. Risks of direct exposure seem to be comparable to those of natural aggregates when blended up to a 15% level. To assess the actual leaching risk, site-specific parameters, as well as knowledge of road configuration and design, are needed to estimate the likely concentrations at the compliance point.

In terms of money, recycled product is worth the challenges faced. In nations where waste incineration is a developed industry, the use of bottom ashes from MSW incineration as an amendment is possible for peatland forestry roads (e.g., Sweden, Italy, Austria, Estonia). Nonetheless, the European Union is prepared to go further in advancing the objectives of the Circular Economy Action Plan, such as recycling secondary raw materials rather than burning them.

8.2.3.4.3. Perspectives of Fly Ashes from the Power Industry for Forest Road Constructions

Fly ashes (i.e., ash residue) are produced in large quantities by power and heating plants (i.e., ashes made up of tiny particles) produced by burning biomass, such as wood and peat; in Finland, for instance, fly ash production reached 700,000 t in 2017. Wood ashes are traditionally used as a fertilizer for agricultural and forest soils. They can also be used as a binding agent in cement, as a material for building roads, or in new materials when combined with freshwater sediments that are rich in organic matter. Particularly for roads through forests and peatlands with little daily traffic, which are paved with gravel or even surfaced with local soil materials, road construction materials are less strict when it comes to design requirements.

Heavy metals do not leach from asphalt roads because water does not percolate through the structure of the road. Although some research has been done, the use of biomass ashes on forest roads has not yet received much attention. According to Bohrn and Stampfer, ashes increased the bearing capacity of a road structure when compared to one without ashes. It was demonstrated in a different study by Kaakkurivaara and Korpunen using a cost-calculation model that ash structures are viable solutions for forest road rehabilitation. The results of a two-year study on the various characteristics of fluidized bed ashes and wood-based ashes indicate that the ashes' chemical and physical states at application fall within environmentally acceptable bounds. It is possible to use biomass ashes without endangering the environment. However, peat-based ashes, where higher concentrations of arsenic may be present, were not evaluated in leaching studies.

The type of fuels used, their purity, the temperature at which they burn, and the equipment used for gas filtration all affect how many heavy metals are present in the ashes from power plants. Although the use of biomass ashes in the construction of forest road infrastructure is thought to be environmentally safe, heavy metal leaching should be taken into account for reasons of environmental safety. When there is a high concentration of biomass ashes widely dispersed throughout the road and water from these areas flows straight into drainage ditches, there is a greater risk of leaching (Vincevica-Gaile et al., 2021).

8.2.3.4.4. Application of Ashes from Other Industries in Peatland Stabilization

The paper and pulp industries produce a lot of ash and look for effective waste management and mechanism solutions that are both cost-effective and environmentally friendly. In a range of studies, laboratory and field tests have been conducted to evaluate the effectiveness of pulp and paper ashes from technical, environmental, and economic viewpoints. Field monitoring has shown that the hydration causes intricate patterns in the ashes' decreased permeability and stiffness over time. Due to their high calcium and aluminosilicate content, fly ashes from the pulp and paper industry have a self-cementitious quality that significantly improves the strength and volume change properties of foundation soils. Ash can therefore be used as a stabilizer to enhance the engineering performance of hydraulic layers, expansive foundation soils, road subgrades and subbase layers, and similar constructions. These ashes have a low permeability, making them suitable for use as hydraulic barriers. They can also be used to cover landfills and sustainably replace materials that are currently in use.

This makes it possible to conserve natural resources by lowering energy consumption, lowering carbon emissions, and substituting building and construction industry materials like cement. Expanding the use of ashes can also prevent the need for additional landfill space for undesirable waste products. Reducing the amount of ash deposited in tailing dams is largely dependent on the use of ashes in earth structures for transport engineering.

Sewage sludge ashes that remain after sewage sludge is burned are what makeup wastewater treatment waste. If dangerous heavy metals are stabilized and leaching is prevented, it can be utilized as a sustainable building material. Concrete, blocks, mortar, road pavements, geotechnical stabilization material, and ceramics can all produce lightweight aggregate. Newly developed materials have to stay away from unfavorable environments like those with high moisture content and acidity. Sewage sludge ashes don't make good binder materials by themselves, but they can be used as an asphalt filler. Various leaching criteria have been used in experimental works to draw comparisons affecting drinking water, soil quality, hazardous material limits, and abrasion tests, which are required for environmental safety. The use of sewage sludge ashes is limited because they are similar to crushed concrete debris and cement kiln dust.

Palm oil fuel ashes (POFA) are waste from tropical industries in southeast Asian countries, e.g., Malaysia, and vast amounts of this material are generated annually. Research into the characteristics and possible applications of POFA, a pozzolanic material that can replace cement, aerated concrete, and lightweight aggregate concrete, was spurred by the abundance of this solid waste. Because the voids are filled with calcium silicate hydrate (C-S-H) gel through an undisturbed hydration process, the POFA specimens containing crushed cockle shells exhibit high strength. This results in denser and stronger bricks.

The production of POFA keeps rising as a result of the expanding plantation areas of palm oil trees in Malaysia, Indonesia, and Thailand. POFA is considered a non-marketable material, and its simple disposal into ponds or lagoons raises serious environmental concerns that negatively impact the sustainability of peatlands. On the other hand, new research indicates that POFA is high in silica and could be useful as an affordable, long-lasting addition to mortars used in the building and road stabilization industries. By working together, POFA and ground blast furnace slags (GBFS) have been able to address the issue of GBFS's quick setting time and create a new kind of green mortar. POFA is a potential material for use in tropical peatlands to stabilize the soil.

8.2.3.4.5. Geopolymer Composites from Thermally Treated Waste

For the past 15 years, geopolymer composites have been studied as a potential replacement for traditional concrete. The binding materials can be used to create secondary and structural elements with good strength and exceptional durability. They primarily consist of pozzolanic materials and alkaline activators. It must be remembered that geopolymer composites require the addition of an alkaline activator, which changes the procedure into a chemical reaction. Examples of such activators include sodium hydroxide and sodium silicate. As a result, numerous attempts were made to make inexpensive concrete and build pavements using this waste material. The results indicate that oil shale ashes have good pozzolanic properties, making them suitable as a starting material for the creation of geopolymer composites for structural and masonry uses.

The characteristics of the products created by alkali activation depend on the choice of raw materials and how they are processed. In addition to different slags and wastes from the glass industry, fly ash combustion can also be helpful in the creation of geopolymer composites. Additionally, using a variety of alkali activators, rice hulls and biofuel ashes have been successfully used to create alkali-activated materials.

The most significant influences on the formation and properties of geopolymers have been found to be the raw material Si–Al ratio, the water–binder ratio, and the rates at which Si and Al are released from source materials. A three-dimensional crosslinked aluminosilicate network is the primary reaction product in the production of geopolymeric binder, which is made from solid fossil fuel residues, most commonly fly ashes of class F with low calcium content. But the remains of class C, i.e., fly ashes with a high calcium content have also been effectively tested for geopolymeric and alkali-activated binders. The geopolymeric potential of oil shale ashes was found to reach 6–7 MPa in this contribution; as a result, there is great potential for using these ashes to cross peatlands and forest roads (Vincevica-Gaile et al., 2021).

8.2.3.4.6. Some Other Specific Materials Used in Peatland Soil Stabilization

Natural plant fibers are a common waste product in agriculture. In sustainable geotechnics, natural fiber materials have become more popular as a method of simulating soil reinforcement. The hydrophilic and biodegradation properties of these materials cause issues and restrict the use of geotechnical engineering techniques. For stable stabilizations to be effective, it is crucial to comprehend how the material behaves under different subsurface conditions. Under subsurface conditions, the shrink-swell behavior of fibers causes slipping and failures. Engineers' current goals are to modify current ground improvement techniques by incorporating the special benefits of natural fiber application, which include resource abundance, low energy consumption, cost-effectiveness, environmental friendliness, and high potential compared to other proven materials. In tropical nations where industry produces large amounts of natural fiber secondary waste, bamboo, jute, coir, palm, sugar cane bagasse, and other fibers have a high potential to stabilize soil; tropical peatland stabilization would be one of the solutions.

When aluminum salts are used in the coagulation process, water purification and wastewater treatment facilities produce alum sludge as a waste. Up to 5% of the total volume of treated water is made up of wet sludge, which is costly and difficult to treat. Landfills are the primary place where wet sludge is disposed of following dewatering and drying. Large-scale waste management issues can be resolved and a cheap soil stabilizer is produced through the use of alum sludge in the stabilization process. Alum belongs to the group of minerals called aluminum hydroxide and is a part of bauxite, which is the main source of aluminum ore.

Usually used for wastewater treatment, it is difficult to dispose of. The California bearing ratio (CBR), which measures foundation strength, increases with the addition of alum sludge to the soil. Both dry and crushed forms of alum sludge can be applied with specific safety precautions. Road construction-related policymakers can recommend using this soil stabilizer to achieve low-cost stabilization, which may result in significant cost savings during projects involving the construction of new roads. In the field, the soil moisture content may rise or fall due to a variety of circumstances, including consistent load application, variations in the water table, varying environmental conditions, and long-term climate changes. Under these circumstances, stabilized soil containing alum sludge improves the stability of building foundations by favorably altering the CBR of weak, clay-like soil in comparison to soil lacking alum sludge. The findings show that an increasing proportion of alum sludge was added at the ideal rate of 8%. It is an affordable alternative to cement and other stabilizers, and it provides a sustainable waste management solution.

Various EU projects, including Wetlands, Algae, Biogas (WAB), Reviving Baltic Resilience (RBR), CONTRA (Baltic Beach Wrack—Conversion of a Nuisance to a Resource), and COST (European Cooperation in Science and Technology) actions like Mining the European Anthroposphere (MINEA), Marine Biotechnology, Protection, Resilience, and Rehabilitation of Degraded Environment, mention combinations of secondary raw material amendments. This approach combines two waste streams, organic and industrial, to create added-value products that will be proposed to businesses and municipalities through a variety of stakeholder channels. Enhancing the dynamics of the waste streams within the system and addressing outstanding issues related to secondary raw materials from municipal activities and energy are critical questions for improving the life cycle of primary products and paving the way for an advanced future of extended producers' responsibility.

When it comes to selecting materials and amendments for weak soil stabilization in the construction industry, including peatlands, the geotechnical engineering community is under pressure to adopt more sustainable practices. To fulfill long-term sustainability objectives, secondary raw (waste) materials are currently meant to replace conventionally manufactured materials.

A robust infrastructure network spanning greater distances is necessary to support the expanding global economy, but economic growth must be balanced with considerations for the environment and the circular economy. If road infrastructure is to be developed and expanded in areas covered by geotechnically challenging substrates like peat, a significant amount of resources are required; consequently, the availability of industrial waste materials and byproducts as secondary resources is crucial, particularly given the possibility of resource shortages in the future. These incentives are coupled with global initiatives like the "Green Deal" and the "EU Industrial Strategy," as well as advancements in the circular economy from both theoretical and practical viewpoints. Utilizing secondary raw materials helps reduce waste and lessens the reliance on raw material supply chains for the construction of infrastructure.

Application and implementation should only be authorized following thorough evaluations of the environmental effects, including leaching and geotechnical bearing capacities. This will save a significant amount of energy and resources, supporting the creative recycling sectors and paving the way for a sustainable future.

8.3. LIFE CYCLE ANALYSIS OF SOIL STABILIZATION METHODS

Soil stabilization is an essential step in the construction of new roads to improve the physical characteristics of soil and make it more suitable for supporting road structures. Reducing plasticity, minimizing settlement, and enhancing load-bearing capacity are the main objectives. For soil stabilization, a variety of techniques are used, such as chemical, biological, and mechanical techniques. Life Cycle Assessment (LCA) can be used to evaluate the life cycle implications of each method.

8.3.1. Life Cycle Assessment Overview

The methodical process of life cycle assessment, or LCA, is employed to assess the environmental effects of a product's extraction of raw materials, manufacture, use, and disposal. When it comes to road construction soil stabilization techniques, LCA aids in determining possible environmental effects at every stage:

- **Raw Material Extraction:** During this stage, supplies such as cement, lime, and other additives that are required for stabilization are sourced. Because of resource depletion and land degradation, the extraction processes may have a major negative influence on the environment.
- **Production:** The creation of stabilizing agents and their delivery to the location are included in this stage. This phase frequently increases energy use and GHG emissions.
- **Construction:** Compaction and preparation of the stabilized soil for road layering are done during construction. The equipment used in this stage could produce emissions and noise pollution.
- **Use Phase:** The stabilized soil's performance during the road's operational life has an impact on longevity and maintenance requirements. Over time, fewer repairs and lower emissions can result from a well-stabilized road because of lower maintenance costs.
- **End-of-Life:** Roads may need to be completely rebuilt or renovated when their useful lives are coming to an end. Evaluating overall sustainability also involves considering how stabilization materials are disposed of or recycled.

Think about things like energy usage, carbon footprint, chemical use, stability affect longevity, and overall environmental impact when comparing various soil stabilization techniques. It's crucial to take into account each method's social and economic benefits, including community acceptability and cost-effectiveness.

Keep in mind that LCA is a thorough and data-intensive analysis, so it's critical to collect precise data and employ approved LCA techniques (e.g., ISO 14040, and ISO 14044) to guarantee the accuracy of your findings. Throughout the study, speaking with professionals in the fields of sustainable construction and LCA can offer insightful advice.

8.3.2. Future Directions in Soil Stabilization LCA

Future LCA evaluations of soil stabilization techniques will probably concentrate on incorporating more environmentally friendly practices into established processes.

- Emphasis on using recycled materials as stabilizing agents.
- Development of bio-based stabilizers that minimize ecological footprints.
- Enhanced modeling techniques that provide better predictions regarding long-term performance and environmental impact reductions.

In short, carrying out a LCA on different soil stabilization techniques yields important information about their sustainability profiles at every stage of their life cycles, from raw material extraction to construction and final disposal. This information enables stakeholders to make well-informed decisions about which techniques best support sustainability objectives (Vincevica-Gaile et al., 2021).

REFERENCES

1. Afrin, H. (2017). A review on different types soil stabilization techniques. *International Journal of Transportation Engineering and Technology, 3*, 19.
2. Al-Khafaji, R., Dulaimi, A., Jafer, H., Mashaan, N. S., Qaidi, S., Obaid, Z. S., & Jwaida, Z. (2023). Stabilization of soft soil by a sustainable binder comprising ground granulated blast slag (GGBS) and cement kiln dust (CKD). *Recycling, 8*, 10.
3. Al-Omari, R. R., & Hamodi, F. J. (1991). Swelling resistant geogrid—A new approach for the treatment of expansive soils. *Geotextiles and Geomembranes, 10*(4), 295–317. https://doi.org/10.1016/0266-1144(91)90008-K.
4. Ashango, A. A., & Patra, N. R. (2016). Behavior of expansive soil treated with steel slag, rice husk ash, and lime. *Journal of Materials in Civil Engineering, 28*(7), 06016008. https://doi.org/10.1061/(ASCE)MT.1943-5533.0001547.
5. Bloch, K. (2024). Strengthening road soils using the stabilization method. *Bachelor's Thesis*, SGGW—Faculty of Civil and Environmental Engineering, Warszawa, Poland.
6. Cabeza, L. F., Rincón, L., Vilariño, V., Pérez, G., & Castell, A. (2014). Life cycle assessment (LCA) and life cycle energy analysis (LCEA) of buildings and the building sector: A review. *Renewable and Sustainable Energy Reviews, 29*, 394–416. https://doi.org/10.1016/j.rser.2013.08.037.
7. Chen, S., Yu, S., Kong, L., et al. (2006). Wrapping method for middle expansive soil embankment and its experimental verification. *Chinese Journal of Rock Mechanical Engineering, 25*(9), 1777–1783.
8. Correia, A. G., Winter, M. G., & Puppala, A. J. (2016). A review of sustainable approaches in transport infrastructure geotechnics. *Transportation Geotechnics, 7*, 21–28. https://doi.org/10.1016/j.trgeo.2016.03.003.
9. Ćwiąkała, M., Greinert, A., Kołodziejczyk, U., Widuch, A., & Wysokowski, A. (2016). *Magazyn Autostrady*. June, 84–93. Gajewska, B., Kraszewski, C., & Rafalski, L. (2018). Significance of cement-stabilized soil grain size distribution in determining the relationship between strength and resilient modulus. *Road Materials and Pavement Design, 19*, 1692–1701.
10. Cyske, W., & Kluska, I. (2007). Comparison of properties of selected binders for soil stabilization at A1 building site. *Roads Bridges-Drog. I Mosty, 6*, 5–18.

11. European Commission. (2021). *Environment, Green Growth, Raw Materials*. Retrieved from: https://ec.europa.eu/environment/green-growth/raw-materials/index_en.htm (accessed on 5 December 2024).
12. Farzad, R. M., Bahman, J. N., & Taghi, R. M. (2021). Laboratory investigation of the effect of "NICOFLOK" polymer on the compressive and tensile strength of desert and coastal sand at the pavement layers. *Amirkabir Journal of Civil Engineering, 53*, 733–748.
13. Gan, W., Zhang, Y., Xu, J., Yang, R., Xiao, A., & Hu, X. (2023). Spatial distribution of soil heavy metal concentrations in road-neighboring areas using UAV-based hyperspectral remote sensing and GIS technology. *Sustainability, 15*, 10043.
14. Gangolells, M., Casals, M., Gassó, S., Forcada, N., Roca, X., & Fuertes, A. (2009). A methodology for predicting the severity of environmental impacts related to the construction process of residential buildings. *Building and Environment, 44*(3), 558–571. https://doi.org/10.1016/j.buildenv.2008.05.001.
15. Hashim, I. H., Badawy, R. M., & Heneash, U. (2023). Impact of pavement defects on traffic operational performance. *Sustainability, 15*, 8293.
16. Jin, J., Tan, Y. Q., Liu, R. H., Zheng, J., & Zhang, J. (2019). Synergy effect of attapulgite, rubber, and diatomite on organic montmorillonite-modified asphalt. *Journal of Materials in Civil Engineering, 31*(2), 04018388. https://doi.org/10.1061/(ASCE)MT.1943-5533.0002601.
17. Johansen, M. D., Aker, P., Klanderud, K., Olsen, S. L., & Skrindo, A. B. (2017). Restoration of peatland by spontaneous revegetation after road construction. *Applied Vegetation Science, 20*, 631–640.
18. Joosten, H. (2016). Peatlands across the globe. In *Peatland Restoration and Ecosystem Services: Science, Policy and Practice* (pp. 19–43). Cambridge University Press.
19. Kenley, R., & Harfield, T. (2011). Greening procurement of infrastructure construction: Optimizing mass haul operation to reduce greenhouse gas emissions. In *Proceedings of the CIB W78-W102 International Conference*, October, Sophia Antipolis, France.
20. Koś, K. (2016). Stabilization of silty soils using cement with an ion-exchanger. *Science Review—Engineering and Environmental Science, 71*, 13–20.
21. Kukiełka, J., & Bańkowski, W. (2019). The experimental study of mineral-cement-emulsion mixtures with rubber powder addition. *Construction and Building Materials, 226*, 759–766.

22. Lei, B., Fu, P., Wang, X., Ni, W., Zhang, S., Wang, X., Shi, J., & Xu, M. (2023). The hydration mechanisms of co-stabilization saline soils by using multiple solid wastes. *Processes, 11*, 2679.
23. Li, X. M., Kong, L. W., Guo, A. G., et al. (2014). Experimental research on shear strength of expansive soil under wetting-drying cycles based on wrapping method. *Rock and Soil Mechanics, 35*(3), 675–682.
24. Liu, Y., Su, Y., Namdar, A., Zhou, G., She, Y., & Yang, Q. (2019). Utilization of cementitious material from residual rice husk ash and lime in stabilization of expansive soil. *Advances in Civil Engineering, 2019*, 5205276. https://doi.org/10.1155/2019/5205276.
25. Miturski, M., Głuchowski, A., & Sas, W. (2021). Influence of dispersed reinforcement on mechanical properties of stabilized soil. *Materials, 14*, 5982.
26. Miturski, M., Sas, W., Radzevičius, A., Šadzevičius, R., Skominas, R., Stelmaszczyk, M., & Głuchowski, A. (2021). Effect of dispersed reinforcement on ultrasonic pulse velocity in stabilized soil. *Materials, 14*, 6951.
27. Moreno-Navarro, F., Sierra, F. J., Sol-Sánchez, M., Rubio-Gámez, M. C., Castillo, M., & Estévez, E. (2020). High-performance sustainable asphalt mixtures for high-volume traffic roads in severe climates. *Sustainability, 12*, 8765.
28. National Technology Approval No IBDiM-KOT-2020/0466 Issue 1. (2024). Chemical binders in use for soil stabilization in road construction substructure. Retrieved from: https://stabimax.pl/cms.files/mix_file/597786,KOT_0469_Spoiwa_chemiczne_SHJ_67.pdf (accessed on 5 December 2024).
29. O'Kelly, B. C., & Pichan, S. P. (2013). Effects of decomposition on the compressibility of fibrous peat—A review. *Geomechanics and Geoengineering, 8*, 286–296.
30. Olchawa, A., Rycharska, J., & Piechowicz, K. (2020). *Design Solutions and Technologies for Agricultural Road Surfaces*. Wydawnictwo ITP: Falenty, Poland. ISBN 9788365426444.
31. Pannier, M. L., Schalbart, P., & Peuportier, B. (2018). Comprehensive assessment of sensitivity analysis methods for the identification of influential factors in building life cycle assessment. *Journal of Cleaner Production, 199*, 466–480. https://doi.org/10.1016/j.jclepro.2018.07.070.
32. Petry, T. M., & Little, D. N. (2002). Review of stabilization of clays and expansive soils in pavements and lightly loaded structures—History, practice, and future. *Journal of Materials in Civil Engineering, 14*(6), 447–460. https://doi.org/10.1061/(ASCE)0899-1561(2002)14:6(447).

33. Piechowicz, K. (2016). Soil stabilization for base of the rural earthroads: Selection of optimal mixtures for stabilization. *Water, Environment, and Rural Areas, 16*, 57–71.

34. Sánchez, M., Wang, D., Briaud, J. L., & Douglas, C. (2014). Typical geomechanical problems associated with railroads on shrink-swell soils. *Transportation Geotechnics, 1*(4), 257–274. https://doi.org/10.1016/j.trgeo.2014.07.002.

35. Scope, C., Vogel, M., & Guenther, E. (2021). Greener, cheaper, or more sustainable: Reviewing sustainability assessments of maintenance strategies of concrete structures. *Sustainable Production and Consumption, 26*, 838–858.

36. Sharifah Zaliha, S. Z., Kamarudin, H., Mustafa Al Bakri, A. M., Binhussain, M., & Siti Salwa, M. S. (2013). Review on soil stabilization techniques. *Australian Journal of Basic and Applied Sciences, 7*, 258–265.

37. Shi, B., Jiang, H., Liu, Z., & Fang, H. Y. (2002). Engineering geological characteristics of expansive soils in China. *Engineering Geology, 67*(1, 2), 63–71. https://doi.org/10.1016/s0013-7952(02)00145-x.

38. Szajna, W. (2021). Chemical stabilization of problematic soils. *Przegląd Budowlany, 92*, 38–43.

39. Teijón-López-Zuazo, E., Vega-Zamanillo, Á., Calzada-Pérez, M. Á., & Robles-Miguel, Á. (2020). Use of recycled aggregates made from construction and demolition waste in sustainable road base layers. *Sustainability, 12*, 6663.

40. Vincevica-Gaile, Z., Teppand, T., Kriipsalu, M., Krievans, M., Jani, Y., Klavins, M., Hendroko Setyobudi, R., Grinfelde, I., Rudovica, V., Tamm, T., et al. (2021). Towards sustainable soil stabilization in peatlands: Secondary raw materials as an alternative. *Sustainability, 13*(12), 6726. https://doi.org/10.3390/su13126726.

41. Zheng, J. L., Zhang, R., & Yang, H. P. (2009). Highway subgrade construction in expansive soil areas. *Journal of Materials in Civil Engineering, 21*(4), 154–162. https://doi.org/10.1061/(ASCE)0899-1561(2009)21:4(154).

CHAPTER 9

IMPLEMENTATION OF STABILIZATION PROJECTS

Contents

INTRODUCTION

A soil's physical characteristics can be changed through the process of soil stabilization to produce long-term, permanent strength gains. Increasing a soil's overall bearing capacity and shear strength is how stabilization is achieved. As soon as it stabilizes, a solid monolith forms, reducing permeability and the potential for shrinkage and swelling as well as the damaging effects of freeze-thaw cycles. A soil's capacity to change volume in response to changes in moisture content is known as its shrink/swell potential. A home, building, or road could easily sustain significant damage from an expansive soil's 10% expansion, which is a drastic volume change. Enhancing in-situ, or natural state, soils can reduce the need for costly remove-and-replace procedures. This is achieved through soil stabilization.

To regulate the engineering properties of a soil, like its moisture content, soils that serve as the structural foundation for roads, building pads, or parking lots are frequently chemically treated. Portland cement, lime, and products derived from lime are some of the chemicals used to stabilize soil. For these substances to create long-lasting bonds between soil particles, pozzolanic reactions are necessary. To ensure that there is sufficient material to stabilize the soil permanently, testing before the project is necessary. The following layers will be thinner and significantly less expensive if the stabilized soil layer is included in the pavement's structural design. When construction, design, and material considerations are given due consideration, lime-stabilized soils outperform non-stabilized soils.

9.1. DESIGN PRINCIPLES FOR SOIL STABILIZATION IN ROAD CONSTRUCTION

One way to define soil stabilization is as a process that combines different techniques for enhancing the stability and strength of a particular soil mass with additional engineering and physical attributes. It is typically used in situations where the soil that is available for construction is unsuitable for the intended purpose. Stabilization results in improved engineering performance and a decreased likelihood of unfavorable engineering behavior, such as washing collapse, etc., taking place by boosting the soil's shear strength, enhancing its ability to support lead, and managing the shrink-swell characteristics. Compaction, pre-consolidation, drainage, and numerous other procedures are all part of the stabilization process. Pure sands and pure clays behave differently in the field, so a soil mass's composition is the primary criterion for stabilization. The strength and hardness of soil are attributed to sand grains larger than 75 microns. However, these particles lack cohesiveness and binding power among their grains, making them easily washed away with relatively unstable slopes. In contrast, clay soils with particles smaller than 75 microns have sufficient binding power between their grains but lack shear strength, especially when saturated. Water, therefore, causes problems for clay soils but gives sandy soils a sense of cohesiveness.

When these two different types of soil are mixed in the right amounts, the result is typically soil with much more superior engineering particles. It is possible to stabilize soil without adding any admixtures at all or to add multiple ones, like cement, lime-pozzolana, lime, etc. To prepare the soil for the intended construction work, it might also entail the use of geotextile or be reinforced with strips. While enhancing the natural soil for building air fields and highways is the primary goal of soil stabilization, it is also employed to modify the permeability and compressibility of the soil mass in earth structures to regulate the grading of soils and aggregates for building air field bases and sub-bases, parking lots, site development projects, and numerous other scenarios where the subsoils are unsuitable for construction. A broad variety of sub-grade materials, from expansive clays to granular materials, can be treated with stabilization (Arora, 2011; Ibetehaj et al., 2014).

9.1.1. Design Factors of Soil Stabilization

Since various factors and variables influence different soil stabilization techniques, it is difficult to identify a single governing principle that applies to all of them. It is widely acknowledged, though, that a few factors should be taken into account before using any kind of soil stabilization, including:

- Evaluate the properties of the given soil type – As previously mentioned, the composition of any soil is the primary criterion for stabilization. A particular soil's engineering qualities are mostly determined by its composition. The right technique to use for stabilizing the soil can be found by comprehending and assessing the given soil's composition and, consequently, its engineering properties.
- Decide the most suitable, effective and economical method of soil stabilization for supplementing the lacking properties – The stabilization technique for clay soils might not work well on sandy soils because of their different engineering characteristics. Even in the extremely unlikely event that it is, it might not be cost-effective in both scenarios. As such, careful planning and consideration must be made when selecting the method or methods that will work best for the stabilization process.

- Design the soil mix with stability and durability values.
- Considering the construction procedure by adequately compacting the stabilized layers.

9.1.2. Classification of Soil Stabilization

For the addition of certain additives, soil stabilization process may be roughly grouped into two:

- Stabilization of existing soil without any additives.
- Stabilization of existing soil with the use of additives – One excellent example of the second, which increases the soil's natural shear strength, is mechanical stabilization. Other examples include bitumen, cement, and lime stabilization. While drainage and compaction serve as excellent examples for the first.

9.1.2.1. Types of Additives

A variety of chemicals, such as cementing agents, modifiers, water retaining agents, water retarding agents, and waterproofing agents, are among the different kinds of additives. Each of these additives exhibits unique behavior from the others, making them each suitable for a specific purpose and having certain limitations. Lime, lime-pozzolana, sodium silicate, and Portland cement are a few examples of cementing agents. Many states have made extensive use of Portland cement to stabilize natural sub-grade soils and improve existing graveled roads.

When free lime reacts chemically with the soil's silica-alumina content, hydrated lime can also act as a cementing agent. In granular materials and lean clays as opposed to expansive or fat clays, hydrated lime works best. For optimal hydration, a comparatively small quantity is usually needed. Cementing materials are expensive, so when they are used for soil stabilization, the soil is only slightly altered rather than actually cementing. This is one of the main drawbacks of using cementing materials. Portland cement, lime, and bitumen are other frequently used modifiers. Very little cement and lime will alter the water film on the soil particles, slightly change the clay minerals, and lower the soil's plasticity index. Bituminous materials, which coat the soil or aggregate grains to slow down and sometimes completely stop the absorption of moisture, are the most effective water-proofing materials. By forcing off the bitumen's volatile components just before compaction, bituminous stabilization works best on sandy soils or low-quality base course materials (Table 9.1).

Table 9.1. The Summary of Additives Used in Different Processes of Soil Stabilization

STABILIZATION	ADDITIVES USED
MECHANICAL	**NONE**
Cementing Agents	▪ Cement ▪ Lime ▪ Lime-pozzolona ▪ Sodium silicate
Modifiers	▪ Cement ▪ Lime ▪ Bitumen
Water proofing agents	▪ Bitumen ▪ Membranes
WATER RETARDING AGENTS	ORGANIC CATONIC COMPOUNDS
Water retaining agents	▪ Calcium chloride ▪ Sodium chloride
Miscellaneous chemicals	▪ Resin ▪ Calcium acrylate ▪ Sulphite lignin

Source: https://www.researchgate.net/publication/342448451_A_REVIEW_OF_THE_PRINCIPLES_AND_METHODS_OF_SOIL_STABILIZATION.

9.1.3. Factors Affecting Mechanical Stabilization

A minimum of two or more natural soil types are combined in the mechanical stabilization process to alter the gradation of the soil and enhance its overall characteristics. This technique aims to integrate the soil mixture's component parts' engineering qualities. By combining soils with varying granular sizes and thoroughly compacting the mixture, the goal is to decrease the void ratio by bridging the spaces between larger granular soil properties with finer soil particles. Granular stabilization is another name for this technique of stabilizing soil. Compaction guarantees a lower void ratio, which enhances soil strength metrics like cohesion (C) and angle of internal friction (Ø).

Concerning mechanical stabilization, these soils can roughly be grouped into two categories, namely:

- **Aggregates:** Here, soils are defined as having an average particle size larger than 75 microns (75M). Strong, well-graded, and somewhat angular sand and gravel particles make up aggregates, which act as a soil's skeleton by creating internal friction and incompressibility.

- **Binders:** These are soils with grain sizes averaging less than 75 microns. For the soil, which is mostly made up of clays and silts, they essentially give it cohesiveness, plasticity, and imperviousness.

To create the appropriate gradation of the mixed soil with both internal frictions supplied by the aggregates and cohesion supplied by the binders, these two soil categories are thoroughly mixed, or "blended," together. Compaction is then used to lower the void ratio of this mixture. Well-graded soils yield easily-achievable soil compaction because they typically have high uniformity coefficients (Cu), typically greater than 15 (>15), and coefficients of curvature (Co), typically between 1 and 3. Certain soil deposits have been known to "show-skip-grading" on some highway projects. This occurs when the particle size distribution curve has a horizontal section at a specific location, indicating that there are no particles of a particular size present. By adding the missing size particles—a process known as blending—a small-scale application of mechanical stabilization is seen here. To facilitate easy compaction using standard road rollers, cohesive and cohesionless soils—which are typically graded uniformly—are occasionally combined. Compaction is typically performed near the optimal moisture content for cohesive and moderately cohesive soils to ensure dry density; for cohesionless soils, vibration is used when the induced frequency (f) is close to the natural frequency of the host soil material.

For shallower soil layers, vibro-compaction, heavy weight compaction, and blast densification are used as part of the dynamic compaction process. For deeper soil layers, rollers, tampers, and other mechanical tools are used. Vibro-compaction is the process of vibrating a probe into the ground and then retracting it after it has densified the nearby soils. If there is less than 12–15% silt and less than 3% clay, it works best. The vibroflot and the terra probe are two popular vibratory probes. A vibratory pile hammer fastened to a steel pipe pile makes up a terra probe, whereas water jets and vibrators are found in a vibroflot. A specialized crane is used in heavy-weight compaction to raise and lower masses weighing between 5 and 30 tons to a height of 12 to 30 meters. While blast densification uses a series of drilled borings with explosives placed in them that, when detonated, release shock waves that densify the surrounding soils, dropping these weights multiple times at a specific location ensures compaction of the mixed soils.

Factors affecting mechanical stabilization are:

- **Strength of Aggregate Used:** The stability of the mixed soil increases with the strength of the aggregates used, though well-planned and compacted mixtures can occasionally make up for the aggregate's relatively low strength.

- **Basic Mineral Composition:** Regarding the stability of the mixture, the basic mineral composition of the mixed soils is crucial. Weather-resistant minerals typically result in mixtures with higher mechanical stability; for example, hydration and dehydration can cause significant volume changes in sodium sulfates and sodium carbonates, which significantly reduce stabilization.
- **Gradation of Mixture:** Greater mechanical stability can only be achieved by carefully designing the soil mixture so that the finer binder particles efficiently fill the pore spaces left by the coarser grains or aggregates.

The effective design of a soil mixture is guided by Fuller's law which states that the granular soil mass attains a very high density after compaction if the distribution of particle size is as follows:

$$\text{Percentage of particle size passing any sieve} = 100 \times \sqrt{D/D_{max}} \quad (1)$$

where; D = opening of a particular size; D_{max} = size of the largest particle.
It is necessary to have a greater proportion of the material passing a 75-micron sieve than that given by Fuller's equation with an ideal mixture containing about 25% binder.

- **Plasticity Characteristics of the Soil Mixture:** These are determined by its binder, which is typically a silty or clay soil with a high plasticity index and liquid limit. This gives the soil more cohesiveness and the ability to retain moisture, acting as a good barrier against surface water traveling downward. The requirements for plasticity, however, vary slightly depending on how the soil moisture is intended to be used. For example, soils intended for the base course have different requirements for plasticity than soils intended for surfacing. Base course soils are supposed to have lower plasticity levels to prevent excessive water buildup and subsequent strength loss while surfacing soils are expected to have higher levels of plasticity.
- **Adequate Compaction of Soil Moisture:** A well-functioning soil mixture may become unstable if it is not compacted thoroughly since the level of compaction attained in the field determines the mechanical stability of the soil mass.

One of the simplest techniques for soil stabilization is mechanical stabilization, which is typically used to improve low-bearing-capacity subgrades. However, heavy-weight compaction, which is used in deep soil compaction, can cause significant shockwaves that can damage buildings and other engineering structures. As a result, it should not be used near existing structures (Arora, 2011; Ibetehaj et al., 2014).

9.1.4. Factors Affecting Stabilization with Cements

Cement stabilization involves mixing water and pulverized Portland cement with soil, then compacting the mixture to create a robust material with minimal moisture fluctuations and enhanced strength and durability. So far, cement stabilization has shown to be a successful technique for stabilizing soil in all soil types except for clay. However, the presence of other organic water in the soil weakens the cement stabilization; for example, sulfates render soil unsuitable for cement improvement. There are three categories into which soil cement falls.

- **Normal Soil Cement:** This consists of between 5 and 14% cement by volume, combined with enough soil to create a cement mixture strong enough for building materials. The amount of water utilized ought to be precisely the right amount to meet the cement's hydration needs and render the mixture workable. Sandy soils and other low-plasticity soils are stabilized with regular soil cement, which is robust and resistant to weather.
- **Plastic Soil Cement:** To have a wet consistency at placement that is comparable to a plasticity mortar, this soil cement contains more water than regular soil cement, even though it still contains 5 to 14% cement by volume. It can be installed on uneven or steep slopes where standard road construction equipment has been effectively used to waterproof the interior of canals and reservoirs and shield steep slopes from the erosive power of water.
- **Cement Modified Soil:** This kind of soil cement is semi-hardened and slightly less effective than the other two, with less than 5% of cement by volume. The soil is not cohesively bound because of the small amount of cement; instead, it interacts with the silt and clay fractions, decreasing their attraction for water and, consequently, their swelling properties.

The following factors affect stabilization using soil types of cement:

- **Soil Type:** Though this amount determined should be suitably increased to account for less efficient mixing of soil and cement in the field relative to that of laboratory experiment, the general rule of thumb is that the proportion of cement to be added to a soil is around 5% for sandy soils and 15% for well graded soils. The exact proportion is determined experimentally by performing a compressive strength and durability test. To maintain the ideal moisture content and achieve the necessary strength and durability, the soil cement needs to be well-compacted. The necessary cement proportion must also be determined experimentally, with equal consideration given to both factors. In addition to being an affordable material for sub-bases and road bases, soil cement can also be used to line irrigation canals and pitch river and canal banks.

- **Quality of Cement:** For a given soil, the amount of cement needed is determined by a laboratory test. Well-graded sands typically require about 5%, while uniformly poorly graded sand may require about 9%, non-plastic silts about 10%, and plastic clay about 13%. A durability test with a maximum permitted volume change of 2% is applied to samples for base course. The test consists of 12 cycles of freezing and thawing or 12 cycles of wetting and drying. When a cement content is increased while the water-cement ratio is decreased, the minimum unconfined compressive strength (UCS) can also be used to calculate the amount of cement needed.
- **Quantity and Quality of Water:** To make cement and silt-clay cement workable, they must be properly hydrated with enough water. Water that is used for hydration and is free of harmful salts, alkalis, acids, or organic matter can generally be obtained in quantities that are determined by taking compaction into account.
- **Adequate Mixing, Compaction and Curing:** A non-homogeneous and weak product could result from improperly mixing the soil, water, and cement mixture, at least until hydration begins, at which point all mixing should stop to prevent a loss of strength. This is because improper mixing could have an impact on the material's stability.
- **Use of Additives:** To boost the soil cement's efficacy, use less cement overall, and help stabilize soil that isn't responsive to cement alone, additional additives may occasionally be added.

9.1.5. Factors Affecting Stabilization with Lime

Every soil particle carries an electrical charge, either a net positive charge or a net negative charge, so adding lime to the soil causes a chemical reaction involving the exchange of cations. Since ions are only loosely held on the surface of soil particles in nature, they are easily replaced by other ions. Water is similarly drawn to soil particles, but because it is bound to exchangeable ions, a process called "adsorption" replaces them. For example, when lime is added to clay soils, the absorbed water layer experiences an exchange of cations, and the soil's plasticity decreases. This results in a more friable clay than the original clay soil, which makes it a more suitable material for construction purposes, such as subgrade or base course for pavements, on embankment slopes, and under canal linings on canal slopes. When lime and wet soil interact, base exchange occurs, which causes calcium ions to replace sodium or hydrogen ions and alters the composition of the absorbed layer.

As a result of the increased cation, the double layer becomes depressed; however, due to the high pH of lime, it may also expand. Although the reaction is dependent on the effective concentration of the reactants and hot temperatures, lime chemically reacts with available silica and alumina in soils to form a naturally occurring substance that functions as cement: calcium alumino silicate complexes. Lime stabilization effectively lowers the soil's liquid limit while raising the plastic limit, which lowers the plasticity index. For this reason, it is widely used in stabilizing extremely unstable plastic soils as well as swelling clayey soils. The soil becomes more friable and workable, and its strength is generally increased, partly because of the soil's decreased plastic properties and partly because of the formation of the cementing material. With the addition of lime, the soil's UCS can increase to 60 times its initial strength, with a higher optimum moisture content (OMC) resulting in a maximum density than in untreated soil. The application of lime treatment causes the clay particles to coagulate into silt-sized particles, altering the soil's structure but stabilizing it by lowering its propensity to swell. Lime is typically obtained by burning limestone in kilns; the quality of the lime is determined by the material and the method of production.

There are basically five types of lime, namely:

- High calcium, quick lime (CaO)
- Hydrated, high calcium lime [$Ca(OH)_2$]
- Dolomitic lime ($CaO + MgO$)
- Normal, hydrated dolomite lime [$Ca(OH)_2 + MgO$]
- Pressure, hydrated dolomite lime [$Ca(OH)_2 + MgO_2$]

Though less convenient and safe, quick lime works better as a stabilizer than hydrated lime. More often used, hydrated lime—also referred to as slaked lime—has a higher magnesium content, which reduces its affinity for water and heat produced during mixing. The quantity of lime needed for stabilization ranges from 2 to 10% of the soil, with the amount needed varying from 1 to 3% depending on the soil's plasticity index. The amount of lime needed increases with the plasticity index value. However, it has been noted that lime does not yield satisfactory results and is generally not very effective for sandy soils unless it is thoroughly and uniformly mixed with the soil, usually with the assistance of machinery such as pulve-mixers. Pozzolona is a siliceous substance that, although having cementitious qualities on its own, when combined with water and finely divided forms reacts to form cementitious compounds that can be used as an admixture. Pozzolana can be made by calcining clay, using naturally occurring volcanic ash, or by using industrial waste like fly ash.

Lime-pozzolona is a widely used admixture. The addition of fly ash, a blast furnace byproduct that is typically high in silica and alumina, tends to accelerate the reaction between the lime and the soil. On the other hand, a comparatively large amount of fly ash is needed for sufficient stabilization. The mixture is referred to as a lime fly ash stabilized mixture when fly ash is added. Not all soils have an adequate amount of clay minerals to facilitate the easy addition of lime. Crushed stone, slag, gravels, silts, and sandy soils are a few of the material types that lime pozzolona can thrive in. Lime to pozzolona ratios can range from 1:1 to 1:9, depending on several variables. A mixture's total amount can range from 10 to 25%. For road bases, lime-pozzolona aggregate mixtures offer superior strength. This material behaves more like a semi-rigid pavement and has excellent structural strength in a layer form. It can be applied to the massive fly ash disposal from thermal plants (Arora, 2011; Ibetehaj et al., 2014).

9.1.6. Factors Affecting Stabilization with Grouting

To improve the soil, the grouting method of soil stabilization involves injecting grouts—special fluid-like materials—into the ground in either a suspension or a solution. Typically, grouting is carried out at high pressure. Because clay soils have low permeability, it is not primarily appropriate for them. Instead, it uses stabilizers with high viscosity for soils with high permeability. The technique is basically appropriate for stabilizing soils that cannot be disturbed and buried zones with limited extent, such as a previous stratum beneath a dam. Therefore, grouting helps to decrease void spaces and raise the soil's capacity to support weight. According to Whitaker and Frith (1990), the following are some necessary characteristics to be considered in choosing a grout to be used:

- **Grout Stability:** During its mixing and injection, the grout should not be set prematurely or sediment.
- **Relative Particle Size:** The size of particles suspended in the grout should be able to penetrate joints/fractures and gouge materials.
- **Low Viscosity:** The viscosity of the grouting fluid should be sufficiently low for easy flow through the soil.
- **Strength:** The grout should be which quickly provides strength to the soil on setting.
- **Performance:** The grout should be able to set well and resist chemical attack or erosion by seeping water.

9.1.6.1. Types of Grouting Methods

There are four major methods by which stabilization by grouting may be carried out.

- **Intrusion Grouting:** This type of grouting, sometimes referred to as slurry grouting, involves injecting grouts through pipes to seal joints or fractures in soil (and occasionally in rocks as well). Either the surface or tunnels are used to insert the pipes. This grouting technique is particularly advantageous because it reduces hydraulic conductivity, which is especially useful for cementitious grouts used in foundations.

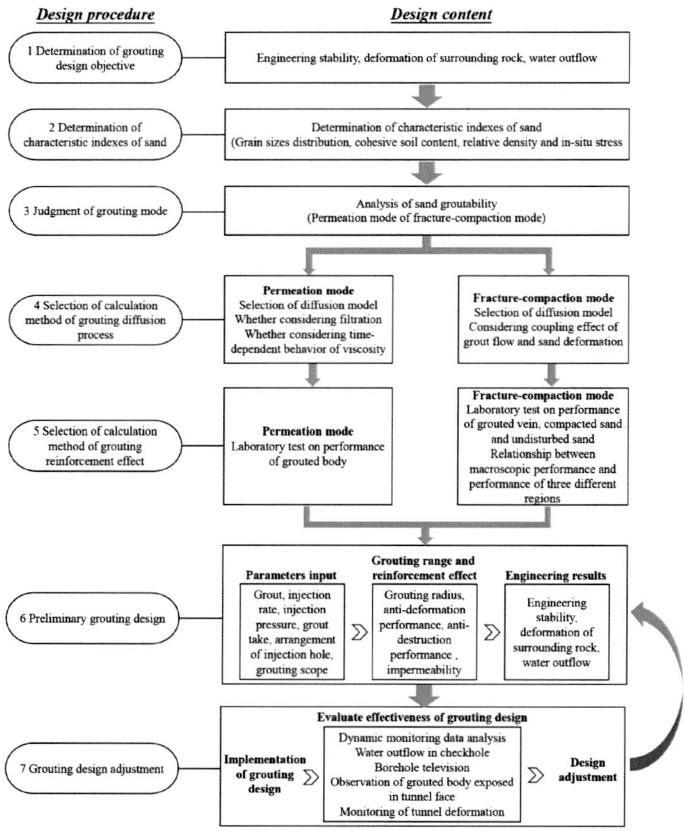

Figure 9.1. *Grouting.*

Source: Li Z, Zhang L, Sun D, Zhang Q, Wang D, Wang L. (2022). Quantitative design method for grouting in sand layers: Practice in Qingdao metro line 2. Processes, 10(5):840. https://doi.org/10.3390/pr10050840.

- **Permeation Grouting:** Here, tiny grouts are injected into the ground to seep into it. When the grouts cure and solidify into a nearly solid mass, they enhance the soil. This type of grouting is sometimes called chemical grouting because it is typically done using chemical grouts. In the end, the treated soil is typically stronger, less compressible, and has a much lower hydraulic conductivity. It is frequently used to stabilize soils before excavating or digging tunnels, as well as to form groundwater barriers (Figure 9.1).
- **Compaction Grouting:** To create a series of inclusions, this grouting technique uses a stiff grout (about a 25 mm slump) that is pumped into the ground under high pressure through a pipe. As this grouting technique raises the structure back into place while also improving the underlying soils, it is typically used to repair structures that have undergone excessive settlement (Figure 9.2).

Figure 9.2. *Sand Grouting: (a) Permeation mode; and (b) Fracture–compaction mode.*

Source: Li Z, Zhang L, Sun D, Zhang Q, Wang D, Wang L. (2022). Quantitative design method for grouting in sand layers: Practice in Qingdao metro line 2. Processes, 10(5):840. https://doi.org/10.3390/pr10050840.

- **Jet Grouting:** Developed in Japan during the 1960s and 1970s, this method of grouting is the most recent. It involves the use of a special pipe with horizontal jets that inject grouts into the soil at high pressure. The pipes are first inserted to the desired depth, and while the injection is taking place, they are raised and rotated to form a column of treated soil. Because of the high pressure associated with the injection, this method works well on a variety of soil types, but it is particularly useful for groundwater control, underpinning, stabilization before tunneling, and several other applications.

9.1.7. Factors Affecting Bituminous Stabilization

Hydrocarbons soluble in carbon dioxide form non-aqueous systems known as bitumens. Bituminous materials like tar, asphaltic bitumen, emulsions, and so forth works particularly well for granular sandy soils and can be used to stabilize soil, especially for sub-grades or road bases. To obtain tars, organic materials like coal must undergo destructive distillation. Natural or refined petroleum bitumens are the main source of materials used to make asphalts. Spread over the soil layer, these bitumen additives serve as both a binder and a water-proofing agent. Asphalts can be cut back with a solvent, like gasoline, and used as an emulsion; however, this requires a longer drying time because asphalts are typically too viscous to be used directly. While somewhat cohesive road bases with these sprays might be slick after rain, they would provide adequate service during dry spells. It is also possible to combine and grind cut-back bitumens with cement-like soils to create stabilized soils. Even in dry conditions, bituminous stabilization can be achieved with any inorganic soil that can be combined with asphalt. While bitumen is needed in amounts ranging from 4 to 7% by weight, trial work has shown that bitumen binds soil particles together in cohesive soils, protecting the soil by filling in its voids and proofing it, helping it to retain low moisture content and increase the bearing capacity.

- **Type Soil:** Sand soils with little to no fines can be stabilized very well with bituminous stabilization. A cohesive soil can be stabilized effectively if its plastic limit is less than about 20% and its liquid limit is less than 40%. However, due to the difficulty in mixing and the substantial amount of asphalt needed, plastic clays cannot be adequately treated. Dry, fine-grained soils in arid areas have dissolved components and values that are resistant to bituminous stabilization.
- **Amount of Asphalt:** Up to a certain point, the amount of asphalt-stabilized soil improves the quality of the soil. On the other hand, an overabundance of asphalt leads to a very fluid mixture that is difficult to compact.
- **Mixing:** The quality of the product improves with more thorough mixing.
- **Compaction:** Compaction type and quantity determine the bitumen soil's dry density. It is contingent on the volatile content as well (Arora, 2011; Ibetehaj et al., 2014).

9.1.8. Factors Affecting Chemical Stabilization

Although it is typically more expensive, chemical stabilization involves stabilizing soil by adding various chemicals. One major benefit is that one can control the soil's setting and curing times. The following are a few substances that have been effectively used for stabilization.

- **Calcium Chloride:** It produces a colloidal reaction when added to soil, changing the properties of soil water. Calcium chloride lessens the amount of moisture that evaporates from the soil because it is hygroscopic and deliquescent. Because it lowers the freezing point of water, it also lessens the likelihood of frost heave. Calcium chloride works wonders as a dust reliever. The method works well for stabilizing silty and clayey soils that lose strength as water content increases because the treated soils do not readily absorb water. The maximum dry density (MDD) is marginally increased by calcium chlorides. The ideal water content is, nevertheless, somewhat less than for the untreated soil. It results in a slight reduction in the soil's strength. Nevertheless, water absorption decreases and treated soil becomes stronger than untreated soil when compacted soil is exposed to water imbibitions. It should be mentioned that the majority of stabilization's advantages depend on the chemical being present in the pore fluid. The advantages disappear as soon as the chemical is removed through leaching. The movement of groundwater has a major impact on how well-treated soils perform.
- **Sodium Chloride:** While the action of sodium chloride and calcium chloride are similar in many ways, sodium chloride has a slightly lower tendency toward moisture. When soil is treated with sodium chloride, the soil's pores begin to crystallize, forming a dense, hard mat with a stabilized surface. The soil's pores fill up and stop more water from evaporating. Moreover, sodium chloride inhibits the propensity for shrinkage cracks to form. It is not advisable to apply sodium chloride directly to soil surfaces; only approximately 1% of the soil's weight is needed.
- **Sodium Silicate:** Soil stabilization has been effectively achieved through the use of sodium silicates and other alkali silicates. The soil is injected with the chemical. When soil reacts with sodium silicate, it becomes stronger. Moreover, it renders the soil impervious. It serves as a dispersing agent as well. It increases the maximum compacted density. The amount of the chemical needed ranges from 0.1 to 0.2% of the soil's weight. Even though it's cheap, it's not appropriate for long-term relationships.
- **Polymers:** Long-chained molecules known as polymers are created

when specific organic compounds known as monomers undergo polymerization. Polymers can be synthetic or natural. Real polymers are called resins. One synthetic polymer that is frequently used is calcium acrylate. Reactions occur when the polymer is added to the soil. Occasionally, a catalyst is used to add monomers to the soil, causing polymerization to occur concurrently with the reaction.

- **Chrome-Lignin:** Lignin is a chemical that is produced as a byproduct when wood is processed to make paper. Black liquor obtained from the production of sulfite paper is used to make chromo-lignin. To create chrome-lignin, potassium or sodium bicarbonate is added to sulfate liquor. It forms a brown gel as it slowly polymerizes. The chemical bonds the soil's particles together through a slow reaction that occurs when it is added. The required percentage of lignin ranges from 5 to 20% by weight. Chrome-lignin is not a permanent solution for stabilizing soil because lignin is soluble in water.
- **Other Chemicals:** A wide variety of other chemicals are also used such as:
  - Soils have been treated with various water proofers, including quaternary ammonium salts, siliconates amines, and alkyl chloro silanes. Coagulating agents, such as calcium chloride and fabric chloride, have been applied to improve the permeability of the soil by increasing electrical attraction and forming a flocculated structure.
  - To promote dispersed structure and increase electrical repulsion, dispersants like sodium hexa-metaphosphate are employed. The soil became more densely packed. Cohesive soils can be stabilized by mixing phosphoric acid with a wetting agent. It forms an insoluble aluminum phosphate when it reacts with clay materials.

9.1.9. Stabilization by Geotextile and Fabrics

Synthetic fiber sheets with permeable properties, such as polyester, polypropylene, polyamide nylon, etc., are known as geotextiles. Whereas geomembranes, which are essentially impermeable polymer sheets or films, can be created by directly spraying asphalt or resin onto the ground, onto geotextiles, or both. They can also be reinforced with linen. Hence, despite being permeable, the geotextiles can be modified to become impermeable geomembranes. Geotextile permeability is like fine sand compared to coarse sand. They are incredibly resilient and strong. Even inhospitable soil conditions have no effect on these.

Over the past 20 years, geotextile use has grown significantly. Geotextiles can be produced in widths up to 10 meters, lengths of rolls up to approximately 600 meters, and thicknesses ranging from 10 to 300 miles (1 mile = 1/100 inch = 0.0254 mm). Geotextiles can be produced in grid, hybrid, non-woven, and woven patterns. Non-woven geotextiles are composed of stable or continuous fibers bonded together at the fiber crossovers using heat, chemical, or mechanical bonding (needle punching). They are widely used in the formation of impermeable barriers as well as filtration, separation, and drainage functions. Continuous monofilament or split film fibers are used to create woven fabrics. The woven materials are primarily utilized for soil reinforcement purposes and offer excellent potential at a reasonable cost. Geotextiles are relatively new synthetic materials that are made from specific polymers using two different methods. They take the shape of nets with varying sizes.

While the basic raw material for both methods is the same, one involves extruded mesh made by a specialized plant, and the other uses special processes that align the molecular chains of polymers to produce material with a higher tensile strength than extruded polymeric mesh. In essence, geogrids are not filters; their primary purpose is reinforcement, and in certain applications, their secondary purpose is separation. Geogrids are frequently used in retaining walls, ground reinforcement, railway track bed stabilization, granular base reinforcement for flexible pavements, slope erosion control, and other applications. In contrast to these geogrid applications, common uses for geotextiles include track foundations, steeper beds, earthen dams and rock fill, increased bearing capacity, protection of river banks, etc. Geogrids and geotextiles are fundamentally different in how they fortify the soil. When it comes to geotextiles, surface friction between the fill and the fabric is what drives the reinforcing function. On the other hand, the reinforcement in the geogrid is accomplished by the fill material positively interlocking into the opening or aperture of the geogrid.

Geotextiles, either as long strips or as cut fabrics, are utilized as soil reinforcement. They are typically made of high-density polyethylene (HDPE) and polypropylene. Anchors at the end can be used to connect them. The anchors, which can help contain soil elements, can be in the shape of rings, loops, or spirals. Mattresses made of prefabricated polymeric systems, called geoweb or geocell mattresses, are used to confine soil or concrete cells. Thick HDPE strips, ranging in width from 50 to 200 mm, are used to construct these systems, and they are sewn or welded at intervals of 50 to 200 m. One possible material for the rhomboidal or hexagonal cells is a geosynthetic honeycomb. They can be folded up for transportation to the job site, where they are then unfolded, put directly on the ground, and filled with dirt. Inside the cells, the material is compressed. They simply need to be compacted once they are filled with dirt. The lining of roads, canals, and footing foundations are all done with these geocells. To create geocomposite, geomembranes, geotextiles, or any similar material that is specifically used for drainage, erosion control, and bank protection must be properly bonded together (Arora, 2011; Ibetehaj et al., 2014).

9.1.10. Functions of Geotextiles

1. **Separation Function:** By keeping gravel from penetrating the clayey sub-grade, geotextile has been used to isolate road subgrade soils from the ground base course, extending the life of the roads. When building a road on wet or weak soil—such as black cotton soils in waterlogged areas—it typically fails quickly because the road crust seeps into the subgrade or soft subgrade soil seeps into the base course gravel. In certain cases, poor sub-grade and base drainage turns into the main factor contributing to road failures. Geotextiles can enhance the current conventional method of construction by lowering the possibility of failure due to its physical functions, which include: (i) Conventional methods have been employed to prevent such pavement failures. Geotextiles serve as an efficient separator, preventing subgrade from penetrating the granular sub-base's voids; (ii) Because geotextiles are highly permeable, water can be quickly removed from the road section while preventing pumping action in the pavement, which is necessary for drainage; and (iii) By increasing the shear strength and CBR value of the sub-grade and lowering the necessary thickness of the pavement crust, the tensile strength of geotextiles helps to spread traffic loads over wider areas and thereby minimizes the possibility of localized stresses in the pavement. This ultimately increases the life of the pavement.
2. **Soil Reinforcement Function:** Tensile members and a geotextile layer work together to create a tensional membrane that serves as reinforcement. Better load distribution is ensured, particularly for heavy weight and high modulus fabrics. Tension is created in the fabric by the deformation under dynamic loads, and the membrane's action serves as reinforcement. Because of the vertical loads, geotextiles maintain a low stress level. When fabric is stretched, the load is distributed over a larger area, reducing its intensity. The underlying soil that geotextiles are placed on is stabilized. It makes the soil stronger in its shear, which raises its bearing capacity and makes it able to support heavier loads. The drainage function of the

geotextile, which allows water to pass through the ballast and out of the roadbed, amplifies this effect even more.

Banks of straightforward and affordable cross-sections can be built to the necessary height thanks to geogrid reinforcement, which offers a workable and long-lasting solution to stability issues. As building progresses, the mesh is layered at predetermined heights and positioned horizontally at the base of the embankment. It extends from the slope's face to the grip length past the calculated lip plane. The soil's shear strength gains a high tensile resistance from Geogrid. The resistance arises from the friction that develops between the soil and the grid structure. Depending on the stability analysis and failure plane location, the geogrid layers may either be continuous throughout the bank's width or curtailed at grip lengths greater than the estimated plane of failure. Providing tensile strength to the soil—which would otherwise be relatively strong in compression and weak in tension—is the main function of geotextiles when they are utilized as soil reinforcement. The tensile strength of the chosen geotextile must be able to be provided at a strain level that is compatible with the soil structure's ability to function. A high tensile stiffness is typically also needed for geotextile soil reinforcement because significant deformations are typically not acceptable in such structures.

Compared with other soil-reinforcing materials, geotextiles have the following advantages:

- They are resistant to all chemical attacks naturally occurring in soils, particularly to alkalis and acids.
- They are found to be stable over a temperature range of –60°C to 100°C. However, the strength is known to reduce at substantially high temperatures.
- They are durable in conditions exposed to prolonged sunlight or wet conditions.
- They are resistant to organic attacks like bacteria and fungi and are not attractive to rats and termites.
- The synthetic fiber mats are known to support vegetation like grasses and expand their growth.
- The geotextiles contain sufficient black carbon which gives them adequate protection during outdoor storage before actual use.

3. **Filtration Function of Geotextiles:** An inverted filter for protection is placed on the ground and covered in concrete blocks with filled open joints during the design of the upstream and downstream protection works of a weir or barrage floor. From bottom to top, layers of fine, coarse, and some stone aggregate make up this inverted filter. The

likelihood of piping is decreased by this arrangement, which permits water seeping from beneath the weir floor to escape without lifting the foundation soil upward. The weir's risk of failing due to piping undermining is decreased by the filter gradation, which permits free passage of seepage water but prevents foundation soil from penetrating and clogging the filter. Geotextile filters, which allow water to flow freely without significantly removing soil particles, are replacing such protective stone aggregate filters because they are more affordable and easier to install, particularly if the filter must be submerged in water. Therefore, good soil retention and transverse permeability are essential qualities for a geotextile filter to have. The following design requirements must be met to use geotextiles as filters: (i) Piping requirements; (ii) Criteria for permeability.

4. **Drainage Function of Geotextiles:** Traditionally, granular drainage filters are placed behind the masonry, which consists of a concrete retaining wall holding up a mass of soil that facilitates water transmission. However, a recent alternative known as fin drain, which consists of a water-conducting filter in the form of a water-conducting core backed with a geotextile filter on either one or both sides, these filters are very prone to clogging and are challenging to construct. Because fin drains have the following benefits over conventional granular drainage, using them behind retaining walls and other straightforward constructions has grown in popularity.
   - They are less susceptible to clogging.
   - They are much easier to install and therefore involve less labor costs.
   - They protect waterproof coating on the back of the wall from drainage during back filling.
   - A wide range of back fill material can be used because of the presence of the water filter sheet on the fin drain.
   - Sometimes the fin drain is pre-assembled with polyethylene on the face against the wall with no other waterproof coating needed to protect the concrete (Arora, 2011; Ibetehaj et al., 2014).

9.2. RECENT EXPERIENCES WITH LIME – FLY ASH STABILIZATION

The long-term advantages of adding a cementitious binder to pavement subgrade soil during reconstruction or new construction have long been understood by pavement engineers. The concept of "perpetual pavement," which is enhanced by "perpetual foundations," has piqued the interest of federal and state highway engineers. When compared to removing and replacing the unstable subgrade soil, stabilizing the soil can save millions of dollars. Stabilizing the subgrade can lead to other pavement layers being thinner when included in the pavement design. In one instance, 2 inches of granular crushed stone base and 5 inches of bituminous base course were removed. In the past, Portland cement has been used to bind non-cohesive, granular, or poorly cohesive soils, while lime alone has been utilized in clay-bearing, highly cohesive soils. Similar to this, the primary purpose of Portland cement is to stabilize an aggregate subbase or base course. Lime and Class F fly ash stabilization can be economically engineered for long-term performance for a low cohesive, silty soil or for reclaiming full-depth asphalt pavement, according to recent investigations and some practical experience. Natural soil with higher amounts of granular material, like sand and gravel, is generally more desirable as pavement subgrade than fine-grained soils with high levels of silt (0.074) and clay (<0.005 mm). Lime stabilization can benefit soils that have a plasticity index (PI) of at least 10 and more than 25% passing through a 74 micron (0.074 mm or 200 mesh) sieve.

The pozzolanic reaction of the lime-fly ash (LFA) combination will significantly improve the bearing capacity and strength of high silt, and low clay subgrades with 25–75% passing the 74 microns and a PI of 10–20, more so than lime alone. The silica and alumina added from the fly ash make up for the lack of reactive silica and alumina caused by the lower clay content. By raising the pH of the soil, the lime helps the pozzolanic formation of calcium silicates and calcium aluminates by solubilizing silica and alumina. Soils having a PI of less than 20 can be stabilized using Portland cement or lime-fly ash. Portland cement is not as effective at breaking down the clay fraction as lime is. It is believed that lower PI soils are better suited for Portland cement, i.e., less than 12, as the low clay content for lime modification allows for faster attainment of the desired strengths. Works published by the U.S. the Transportation Research Board and Army Corps of Engineers acknowledge the value of lime-fly ash for coarser-grained soils with little clay or plastic fines, such as silt.

There are two possible forms of lime: hydrated lime (calcium hydroxide, $Ca(OH)_2$), or quicklime (calcium oxide, CaO). Before supplying the calcium to react with the silica and alumina in the clayey soil and from the addition of coal fly ash, quicklime hydrates with the soil moisture to become hydrated lime, acting as a better drying agent. The main component of lime kiln dust (LKD) is fly ash (8–15%), which comes from the coal used to burn or calcinate the limestone into lime (20–40% calcium oxide). When used in the right amounts, LKD is used for both soil stabilization and modification. It is important to distinguish between soil stabilization and soil modification. Modified soil due to moderate addition rates (i.e., 3% quicklime) results in benefits like swell reduction and drying. This presentation focuses on soil stabilization from higher application rates when long-term freeze-thaw protection strengths are needed (Arora, 2011; Ibetehaj et al., 2014).

9.2.1. Lime – Fly Ash Pozzolanic Reactivity

In order to stabilize soil subgrade and granular base coarse beneath pavements or foundations, lime and coal fly ash are used in a pozzolanic reaction. This paper aims to relate new and recent experiences and information regarding recent investigations at the Carmeuse Technology Center and some pertinent work of others. As described in ASTM C-593, the plastic mortar cube test method can be used to measure or characterize the pozzolanic cementitious strength-gaining reaction between a given fly ash and hydrated lime. Around 73% of the solid mix is composed of graded standard ASTM sand as the aggregate. The dry mixture is composed of 8.9% lime and 17.8% fly ash, a 1:2 ratio. This process is recommended for non-self-cementing Class F fly ash rather than the self-cementing fly ash method, which doesn't require any standard graded sand. The fixed flow method, as opposed to the fixed water/cement ratio method, determines the amount of water added. To achieve a standard water flow or fluidity, more water is added (slump cone test), a user must fall within a specific range following 10 blows on the drop table. Depending on their fineness and surface properties, some fly ashes will require more water to achieve this fluidity. Fly ash's loss on ignition (LOI) is not as important for soil stabilization as it is for concrete, so ASTM C-618's LOI limitations do not necessarily apply. But it's unknown what happens if the LOI is higher than 10. At the Carmeuse Technology Center in Pittsburgh, Pennsylvania, a total of 18 fly ashes—both Class F and Class C—have been tested. An example analysis is in Table 9.2. By using lime to test, all of the fly ashes tested higher than 600 psi. Researchers test four Class C ashes both with and without lime. When lime was added, three of them significantly improved. Without lime, only the Lakeshore C ash in the Cleveland area performed better.

Table 9.2. *Chemical Analysis of Calcined Clay Samples with Pozzolanic Potential*

Chemical Elements	Sample Clays (%)		
	A1	A2	A3
SiO_2	48.13	46.04	38.04
Al_2O_3	41.56	38.71	35.02

Fe_2O_3	3.76	7.99	15.71
K_2O	2.41	2.22	1.63
SO_3	2.14	1.90	1.79
TiO_2	1.72	1.35	1.59
CaO	0.07	0.35	0.71
MnO	0.05	0.07	0.00
ZrO_2	0.03	0.04	0.00
ZnO	0.02	0.00	0.03
SrO	0.01	0.00	0.00
P_2O_5	0.00	1.15	0.00
V_2O_5	0.00	0.12	0.08
Ir_2O_3	0.00	0.06	0.00

Source: Pinheiro VD, Abreu RFd, Alexandre J, Xavier GdC, Marvila MT, de Azevedo ARG. (2024). Pozzolanic potential of calcined clays at medium temperature as supplementary cementitious material. Sustainability, 16(17):7508. https://doi.org/10.3390/su16177508.

9.2.2. Mixture Design and Testing Protocol

9.2.2.1. Moisture-Density Relationship and Lime Demand

The OMC of the soil stabilized mixture and the dosage of lime or LFA blend must be established before conducting strength tests. The lime dosage rate, also known as the lime demand, is calculated using the EadsGrim soil pH test (ASTM D-6276). OMC tests use ASTM standard compaction procedure D-698 to determine the maximum dry density (MDD). To demonstrate how the lime or LFA alters the material handling aspects (i.e., the OMC of the treated and untreated soils should be compared). reduction of plasticity and drying of the soil) and the simplicity of achieving the necessary compaction/density. The addition of lime causes the properties of the soil to change, increasing the amount of moisture required for compaction while decreasing the MDD. Similarly, there is a decrease in the soil's plastic index (PI), indicating a simpler process for compaction (Figure 9.3).

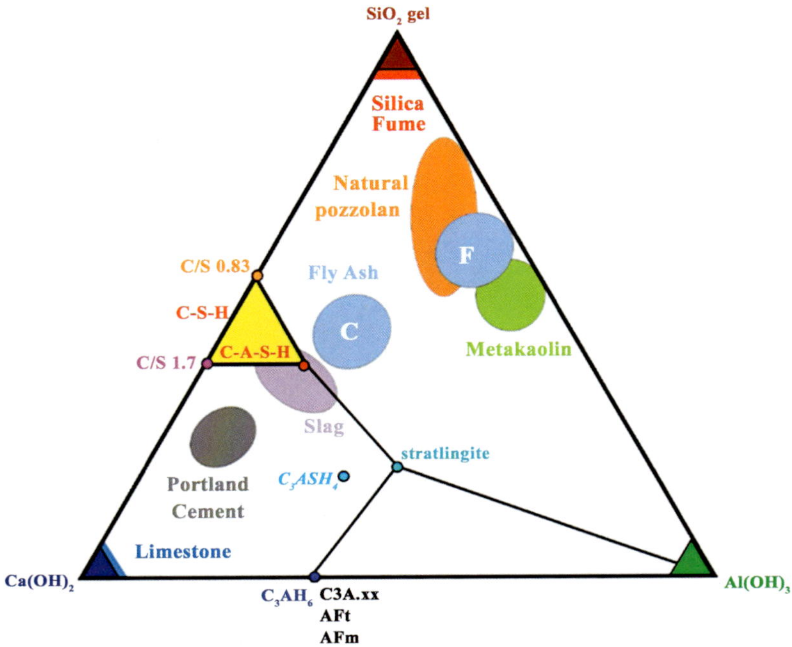

Figure 9.3. *The compositions of the studied samples are plotted within the natural pozzolan field on the Ca-Si-Al ternary diagram, presenting the different average compositions of Portland cement, limestone, and SCMs.*

Source: Sainis T, Kakali G, Pomonis P, Vasilatos C. (2023). A comparative study on the properties of volcanic rocks from the Aegean islands, Greece, for utilization as pozzolanic additives in cement. Materials Proceedings, 15(1):21. https://doi.org/10.3390/materproc2023015021.

9.2.2.2. Strength Gain and Moisture Sensitivity via UCS Testing

UCS tests were determined using ASTM D-5102. Both untreated and treated samples are prepared for strength testing at their respective OMC contents. The curing conditions (time and temperature) are chosen from three techniques:

- Accelerated curing for 3 days at 49–50°C; (120°F);
- Accelerated curing for 7 days at 40°C (104°F)
- Ambient temperature curing for 28 days at 22°C (72°F).

Numerous investigations have revealed that these curing times and conditions ought to produce roughly equal strengths. It is possible to ascertain moisture susceptibility by letting some of the cured UCS cylinders (i.e., following seven days of accelerated temperature curing) to undergo a 24-hour capillary soaking procedure before strength testing. The 4-inch diameter UCS cylinder, which is covered in a wet absorptive cloth, is placed on a porous stone submerged in water as part of the capillary soak protocol. Comprehensive laboratory testing has shown that after the capillary soak, untreated clayey soils usually degrade to a compressive strength of less than 10 psi. Though it may only lose 15–25% of its strength, stabilized soil should still be able to meet target requirements for "various anticipated service conditions." Strength specifications for "different expected service conditions" (i.e., conditions (such as soaking or periodic freeze-thaw cycles) for the application of base, subbase, and subgrade. The required strength ranges from 50 to 200 psi, depending on the application and service condition. The U.S. Army specifies 250 psi for subgrade beneath asphalt and 200 psi for rigid concrete.

9.2.2.3. CBR Testing

After soaking for four days at room temperature, the CBR strength test is carried out using a piston-like penetration force at a particular density and moisture content (often the OMC). Comparing treated and untreated soil, as well as those with and without a curing period (i.e., seven days at 40°C). During the soaking time, a strain gauge measures the expansion or swelling. When measured relative to crushed stone, the CBR is expressed as a percentage. Following a 4-day soak, moisture absorption is measured. The need for stabilizing the soil subgrade and the necessary total pavement thickness above the subgrade are determined using CBR data. Results from other tests, like the resilient modulus stiffness test and the dynamic cone penetrometer (DCP), can be correlated with lab CBR results.

9.2.3. Lime – Fly Ash with Full Depth Reclamation

A predetermined amount of the underlying subgrade and the entire or full-depth pavement section are uniformly crushed, pulverized, or blended to create a stabilized base course. This process is known as full depth reclamation (FDR), and it is a bituminous concrete recycling technique. When the granular base is no thicker than six inches or when the entire base has structurally failed, FDR is especially appropriate. Chemical stabilizing additives can be added in a dry or liquid form before the pavement is ground up and combined. Lime and coal fly ash are examples of dry chemical additives. Few FDR projects utilizing LFA have been documented to date (Figure 9.4).

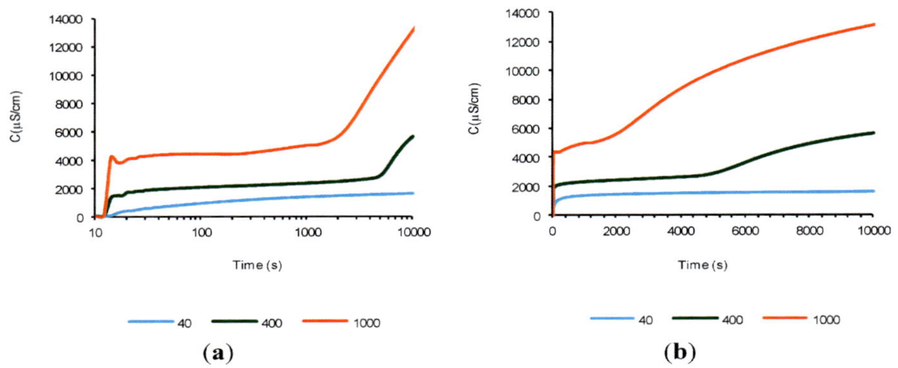

Figure 9.4. *The time evolution of electrical conductivity of aqueous suspensions with several cement amounts (mg), T = 40°C: (a) semi logarithmic scale; and (b) normal scale.*

Source: Velázquez S, Monzó JM, Borrachero MV, Payá J. (2014). Assessment of pozzolanic activity using methods based on the measurement of electrical conductivity of suspensions of Portland cement and Pozzolan. Materials, 7(11):7533-7547. https://doi.org/10.3390/ma7117533.

9.2.4. Stabilizing Black Cotton Soil

Medium to high compressible inorganic clay soil makes up black cotton soil. This soil's characteristic is that it swells when it gets wet due to its high liquid limit and high plasticity when it's dry. Therefore, it is not advisable to build any kind of structure on this land. This paper reports a study that used lime in the form of quick lime and bottom ash to stabilize the soil for growing black cotton. These elements are mixed into the soil in varying amounts, and the results are extracted and used to stabilize the soil in preparation for pavement construction. Due to the characteristics of black cotton soil, it swells during the rainy season and contracts during the summer. Because the soil is more pliable during the rainy season, the building's structure will move upward and downward and sink deeper into the ground. Our goal is to increase the soil's shear strength and bearing capacity by applying bottom ash and lime. It is discovered that pavements built in BC soil areas experience early failures. In particular, excessive unevenness, ruts, waves, and corrugations will form almost immediately following every monsoon session on flexible pavements with heavy traffic, leading to a high maintenance demand and associated costs. It is discovered that the "montmorillonite" mineral, which has highly expansive properties, is present in the black cotton soils. In BC soils, the percentage of the finest particles, or colloids, will reach up to 50%, while the percentage of grains larger than 75 micrometers will fall between 75% and 100%. Due to its high plasticity, compressibility, and sensitivity to changes in moisture, the BC soil is characterized by these characteristics. The primary issues are as follows: If the large mass of soil (lumps) is dry, it is difficult to pulverize it. If it gets wet, it gets too sticky.

There are extreme variations in plasticity and shrinkage, with variations in water content.

- Shrinkage is too much upon drying, this creates excessive cracks.
- BC soils exert high swelling pressure.
- In areas where BC soil is prevalent, hard aggregates may not be available within a short distance.

The best technique for stabilizing soil in British Columbia is to use lime and, if needed, additional additives. Effectiveness in thermal stabilization has also been discovered. Heat treating the soil as a whole, however, is not only expensive but also ineffective. Additionally, there will be a significant need for cement, and the procedure won't be cost-effective. Lime is therefore the ideal substance for blending. The changes that occur in BC soil following lime stabilization are listed below.

- Decrease in plasticity index. It becomes easier to pulverize BC soil lumps.
- The affinity with water decreases. There will be a low volume change.
- The stability of BC soil under soaked conditions will increase.
- The shrinkage limit is increased and there will be no shrinkage after compaction.
- The cement requirement for BC soil with lime is very low when compared with untreated soil.

All these modifications are desirable changes for the stabilization of BC soils for use as a subgrade and sub-base course material. Following are the recommendations regarding the stabilization of low-volume roads:

- The soil subgrade is to be treated with a small portion of lime (about 1% of weight) up to the required depth.
- Over well-compacted soil-lime, a subgrade may be developed. Two steps can be taken in the lime stabilization process: first, precondition the BC soil with 2 to 3% lime, and then, repulverize the altered soil and add the remaining lime. The layer's thickness is determined by the construction site's equipment and design factors.

- A base course can be made with soil lime and a small amount of additional additives. Portland cement could be added in amounts of 3 to 4%. Soil-lime-flyash or soil-lime-heat-treated BC soil can also be used as alternatives to cement.

9.2.5. Stabilization of Desert Sand

Large amounts of desert sand can be found in parts of Rajasthan and a few other Indian locations. Anytime such sands need to be stabilized, it presents an extremely difficult challenge due to a severe water shortage and the lack of other resources. As a result, any recommended stabilization strategy ought to eventually benefit the economy. The particles in desert sand are consistently finely graded and contain a higher proportion of rounded particles.

The sand becomes very unstable as a result. To stabilize such sands, a significant amount of cement will also be needed. Water scarcity makes soil-cement stabilization extremely expensive. Considering these factors, bituminous stabilization is always advised for soil in desert regions. It would be best to use a hot sand-bitumen mixture since it yields a satisfactory mixture. Additionally, the locally accessible Kankar dust is utilized for mix preparation. In this case, cut-back bitumen is determined to be more suitable. Cement is typically used to stabilize sandy soils. The final compacted density and the type and quantity of fines present in sandy soils determine how much cement is needed to stabilize the soil. Sandy soils require 5 to 12% of their weight to be cement. Bitumen emulsion is the most commonly used bituminous material.

There won't be much more water required during mixing because the emulsion already contains 40% water. The bitumen emulsion stabilizes the sand during curing, allowing for extended use. By adding additional materials to improve the gradation and mixing in Kanker Power, the mix's stability can be further increased. Try bituminous emulsion with appropriate additives as well. It is not possible to support structures on the desert sands due to their low bearing capacity and tendency to collapse when wet.

Cement by-pass dust, also known as cement kiln dust, and Portland cement were added in amounts of 2, 4, 8, 10, and 12 by dry weight of soil as stabilizing agents. To measure the engineering properties of the stabilized material, laboratory tests like compaction, unconfined compression, and shear box tests should be carried out. The MDD, UCS, and shear strength parameters (c and φ) must all show significant improvements. Therefore, the compressibility and shear strength properties of desert sands can be enhanced by adding cement and cement by-pass dust.

9.2.6. Optimization of Fly Ash-Based Soil Stabilization

To combat environmental problems brought on by the building and energy sectors, the construction industry is pursuing sustainable resource use. Currently, waste materials recycling and reuse are receiving more attention to support engineering projects. These resources include foamed glass that has been recycled, building and demolition debris, and even leftover food and drink. These kinds of research projects help to recycle waste materials for use in engineering projects. For instance, since 2011, crushed glass and glass fines have been allowed to be used in road pavements in Victoria, Australia. Additionally, because of fly ash's widespread availability and the advantages it offers for waste management and ground improvement, it has garnered a lot of attention lately. Even though coal ash is currently utilized in a variety of ways in several construction applications, its rate of utilization is so much lower than its rate of generation that it accounts for nearly one-fifth of Australia's total waste stream. Furthermore, according to reported statistics, only 44% of the 12 million tons of coal ash produced annually—of which only half is used for useful purposes, primarily as fly ash—is recovered from the dumps.

Depending on the kind of coal burned, fly ash can contain pozzolans, which can trigger chemical reactions and eventually strengthen soils for use in geotechnical applications. The disposal of fly ash, a byproduct of the production of energy from coal, can expose energy companies to significant financial and environmental risks. The literature reports on studies that demonstrate the beneficial effectiveness of using fly ash sustainably. Examining the impact of using fly ash in place of some of the cement in recycled aggregate concrete led to findings about a slight decrease in yield strength and a minor reduction in strength, but the material also showed improvements in flowability, plastic viscosity, and chloride resistance, all of which improved the engineering application's sustainability. By using all of the fly ash that is available in China, for example, savings of up to 150 billion CNY and 560 million tons of carbon dioxide could be achieved. Fly ash use has the potential to significantly reduce social, environmental, and economic effects. The utilization of coal fly ash (CFA) as a sustainable material for the removal of anionic surfactants is made possible by its lower embodied energy when compared to carbon-based surfactants. Additionally, CFA's adsorbent properties can effectively remove high concentrations of sodium dodecyl sulfate (SDS). Fly ash is a sustainable pavement base material that has been shown to be effective in further studies. Reduced waste disposal problems and a decreased need for calcium-based soil stabilization additives like cement can be achieved through greater use of fly ash. It is primarily the pozzolanic reactions that take place between the soil and the calcium-based additive that make fly ash an effective subgrade treatment.

To create calcium silicate hydrate (CSH) and calcium aluminate hydrate (CAH), pozzolanic reactions entail the movement of calcium hydroxide by water and its combination with silicate and aluminate clay minerals. Research has demonstrated how the strength, flexural fatigue, and stiffness of pavement materials are influenced by the calcium and alumina contents. The production of aluminosilicate gels has been demonstrated in the absence of calcium-rich precursors, using fly ash or other silica and alumina-rich precursors. Fly ash cements surfaces, but studies indicate that fly ash by itself cannot sufficiently improve geotechnical properties for use in pavement and foundation design. The literature has, however, noted that fly ash use can be somewhat successful, particularly when paired with other additives. Fly ash and cement treated clay soils resulted in a substantial amount of tobermorite formation, which gave the samples a denser and more stable structure and better strength and geotechnical qualities. In addition, the financial benefits of these additives are increased when comparing pavement structures with these stabilized subgrades to traditional flexible pavements because the asphalt thickness is reduced.

UCS of the material with 7.2% SO_3 was found to be up to 120% higher than that of the admixture with 1.8% SO_3. This study examined the impact of SO_3 content on cement and fly ash stabilized crush stones. With fly ash and rice husk ash (RHA) contents of up to 25% and 12%, respectively, the potential benefits of the admixture were demonstrated by significantly increased UCS and CBR. Enhancement of the soil's plasticity index and compaction limits, as well as the best possible UCS and CBR strength improvement of fly ash and lime, treated samples at 20% fly ash and 18.5% lime. When soluble alumina and silica from clay and fly ash combine with calcium from lime and fly ash in the presence of water, stable CSH and CAH are produced. This increases the soil's long-term strength and enhances its geotechnical qualities. The majority of stabilization work to date has involved the combination of cement and fly ash, despite the fact that numerous other combinations, including fibers and ground granulated blast furnace slag (GGBFS), have also been reported to demonstrate sufficient strength gain.

The sustainability of pavement construction techniques would be increased by straying from conventional ground treatment techniques. The application of enzymes as a stabilizing agent for soil has garnered significant interest ever since they were first used in the building of roads. The mechanism of enzymatic stabilization was not completely understood until recently, although enzymes have been commercially available for the last 40 years. The enzyme bound to a substrate based on its active site, where a reaction takes place easily to convert the enzyme-substrate complex to the product of the reaction; a decrease in the double layer of water of clay from the encapsulation of organic materials;

aggregation of clay particles and reduction in surface area; and microbiologically induced calcite precipitation (MICP) are some of the hypothesized mechanisms for the successful stabilization applications.

By densifying and reducing the affinity for water, enzyme-based soil stabilizers with amylase, lipase, and protease as active ingredients help effectively stabilize clayey soils. This results in a leftward shift in the stabilized material's OMC and an upward shift in its MDD, producing stronger material with less compaction energy. Although not enough to be used as subbase or base course materials, enzymes containing the same essential ingredients can significantly alter the subgrade quality of marginal soils for use as materials for highway embankments. Studies that compare give non-traditional additives based on calcium a higher rating than those that rely on enzymes. Thomas et al. (2018) found that stronger soil can be achieved with alkali-activated ground granulated blast furnace slag (GGBS), which is a more effective treatment option than soil treated with OPC or soil treated with enzymes. Strength improvements can reach up to 11.5 times higher with GGBS (Arora, 2011; Ibetehaj et al., 2014).

REFERENCES

1. Aboud, T. T., Kasa, A. B., & Chik, Z. B. (2007). Stabilization of silty clay soil using chloride compounds. *Journal of Engineering Science and Technology, 2*, 102–103.
2. Arora, K. R. (2011). *Soil Mechanics & Foundation Engineering* (7th ed.). Standard Publisher Distributors.
3. Behzad, K., & Huat, B. B. K. (2008). Peat soil stabilization using ordinary Portland cement, polypropylene fibres and air curing. *Electronic Journal of Geotechnical Engineering, 13*, 1.
4. Birchal, V. S. S., Rocha, S. D. F., & Ciminelli, V. S. T. (2000). The effect of magnesite calcination conditions on magnesia hydration. *Minerals Engineering, 13*(14, 15), 1629–1633.
5. Brown, R. E., & Glenn, A. J. (1976). Vibroflotation & terra-probe comparison. *Journal of the Geotechnical Engineering Division, 102*(GT10), 1059–1072.
6. Ibetehaj, T. J., Mohd, R. T., Zaid, H. M., & Tanveer, A. K. (2014). Soil stabilization using lime: Advantages, disadvantages and proposing a potential alternative. *Research Journal of Applied Sciences, Engineering & Technology*.
7. Whitaker, B. N., & Frith, R. C. (1990). *Tunneling, Design, Stability, and Construction*. The Institute of Mining & Metallurgy.

CHAPTER 10

FUTURE TRENDS AND CHALLENGES IN SOIL STABILIZATION

Contents

INTRODUCTION

Because of the need for sustainable practices, environmental concerns, and technological advancements, the field of soil stabilization is changing quickly. Future developments in soil stabilization will involve incorporating cutting-edge testing methods, novel materials, and ecologically friendly strategies. The use of geosynthetics, such as geotextiles and geomembranes, which improve soil stability and lessen the need for conventional, more resource-intensive methods, is one noteworthy trend. By effectively reinforcing and filtering, geosynthetics reduce environmental impact and enhance load-bearing capacity and erosion control. The application of sustainable additives and bio-based stabilizers is another encouraging development. In place of traditional chemical stabilizers, researchers are looking into natural materials like biochar, organic polymers, and plant-based resins. These bio-based solutions have the potential to be more sustainable and have less of an impact on the environment. Furthermore, the development of nanoscale additives that can improve soil properties in small amounts is made possible by advances in nanotechnology, which lessens the need for large amounts of conventional stabilizers.

Innovations like real-time data collection and remote sensing are revolutionizing testing and monitoring procedures for soil stabilization. Drones and sensors are being used more and more to track soil performance and conditions both during and after stabilization. This data is crucial for enhancing treatment strategies and guaranteeing their long-term efficacy. With the aid of this technology, soil stabilization projects can be managed with greater accuracy and agility. Even with these developments, a number of obstacles still exist. Dealing with soil contamination is a significant obstacle that makes stabilization efforts more difficult and calls for specific remediation methods. Research is still needed to find effective solutions because the interaction between contaminants and stabilization materials can reduce the efficacy of treatments. Significant obstacles also include the unpredictability of soil conditions and the effects of climate change. The behavior and performance of soil can be altered by extreme weather events and shifting environmental conditions, calling for flexible and robust stabilization techniques. Overall, innovation, sustainability, and improved performance are the main themes of the future of soil stabilization. Infrastructure projects can be made stable and long-lasting in an ever-changing environmental landscape by utilizing new materials and technologies and resolving persistent challenges (Goetz, 2011; Marinelli et al., 2018).

10.1. EMERGING TRENDS IN SOIL STABILIZATION RESEARCH

One of the key components required to determine how a construction project will be carried out is soil. Worldwide, there is a growing need for various types of infrastructure development due to rapid urbanization. The viability of the site is a key factor in project selection; in general, good bearing capacity soil is preferred. For the majority of geotechnical projects, the lack of available land makes it impossible to obtain a construction field that satisfies design specifications without modifying the ground. In these conditions, it is imperative to prioritize the subsurface that can withstand construction. The use of weak and problematic soil and its transformation into good soil through the application of any appropriate technology necessary to enhance the behavior of the soil and thus its inherent properties presents challenges. Due to its volume change behavior when interacting with water and the presence of minerals called montmorillonite clay, which are expansive in nature and cover about 12% of the landmass, expansive soil is among the most problematic types of soil. An essential technique for increasing the soil's load-carrying capacity is ground improvement. It consists of many methods that vary depending on the type of soil and other factors. The stabilization method is one of the most efficient ways to improve the strength properties of natural soil by adding material to it. It has a significant role in the quick development of characteristics. Physical, chemical, and biological stabilization techniques can be used to improve the in-situ subsoil's performance and capacity to support weight. Cement stabilization, lime stabilization, flash stabilization, bitumen stabilization, and other methods are frequently employed stabilization techniques.

Improving soil qualities to increase the performance and longevity of construction projects is the main goal of soil stabilization research and application in civil engineering. Future developments in soil stabilization research are influencing resilient and sustainable building techniques as environmental concerns gain prominence and infrastructure demands increase. This thorough investigation explores the most recent developments, techniques, and creative strategies that are revolutionizing soil stabilization (Goetz, 2011; Marinelli et al., 2018).

10.1.1. Advancements in Stabilizing Agents

10.1.1.1. Biopolymers and Bio-Based Stabilizers

The use of biopolymers and bio-based stabilizers, which provide environmentally friendly substitutes for conventional chemical stabilizers, has drawn more attention in recent years. Biopolymers, which can improve soil strength and lessen environmental impact, are environmentally friendly alternatives. Examples of these include chitosan, starch, and cellulose. When compared to synthetic chemicals, these materials have the potential to reduce carbon footprints while also enhancing soil cohesion.

Another area of soil stabilization is nanotechnology, which uses nanoparticles to change the microscopic characteristics of soil. Nanomaterials can greatly increase soil compaction, decrease permeability, and improve overall stability. Examples of these nanomaterials are silica, alumina, and clay-based nanoparticles. Research is still being conducted to better understand the long-term effects of nanoparticle formulations on soil behavior and to optimize their use in soil stabilization.

10.1.1.2. Recycled and Waste Materials

The use of waste and recycled materials in soil stabilization techniques is becoming more popular. The potential of materials like fly ash, slag, and industrial by-products to enhance soil qualities and encourage waste reuse is being investigated. By strengthening the soil, decreasing swelling, and increasing durability, these materials provide a sustainable method of stabilizing the soil that is in line with the ideas of the circular economy (Goetz, 2011; Marinelli et al., 2018).

10.1.2. Innovative Stabilization Techniques

- **Electrokinetic Stabilization:** Applying electric fields to soil to enhance its characteristics is known as electrokinetic stabilization, and it is a relatively new method that has promise for difficult soil situations. This technique improves compaction, lowers soil permeability, and aids in the consolidation of fine-grained soils. Investigations into practical applications in diverse soil types and the optimization of electric field parameters are still in progress.
- **Geosynthetics and Geocomposites:** Geosynthetic materials, such as geotextiles, geomembranes, and geogrids, are essential elements in soil stabilization that are still developing. Recent developments concentrate on enhancing these materials' mechanical qualities and creating geocomposites—Which combine several purposes, including drainage, reinforcement, and separation. Geosynthetics provide creative ways to improve load-bearing capacities, stabilize slopes, and reinforce soils.
- **Biocementation:** Using microorganisms, biocementation is a novel

process that causes calcium carbonate crystals to form in soil. By causing minerals to precipitate and bind soil particles together, this biological process improves the strength and stability of the soil. Numerous microbial species, ideal mineral precipitation conditions, and useful applications in geotechnical engineering and construction are all explored in biocementation research.

- **Mechanical Stabilization:** Using physical techniques to improve soil qualities, such as mechanical compaction and soil mixing, is known as mechanical stabilization. In order to achieve superior soil improvement, recent trends have focused on combining mechanical stabilization with other methods, such as chemical or biological treatments. More accurate control over soil properties and treatment efficacy is now possible thanks to technological and procedural advancements.

10.1.3. Sustainability and Environmental Considerations

- **Life Cycle Assessment:** An increasingly important method for assessing the sustainability of soil stabilization techniques is life cycle assessment, or LCA. The process takes into account how soil stabilization techniques affect the environment at every stage, from production to application and disposal. Researchers and practitioners can make better decisions regarding the environmental effects of different stabilization techniques by evaluating factors like energy consumption, emissions, and resource use.
- **Carbon Footprint Reduction:** A major area of current research is lowering the carbon footprint of soil stabilization techniques. Greener construction practices are being aided by advancements in low-carbon stabilizers, energy-efficient techniques, and the use of renewable materials. Researchers are looking into ways to incorporate carbon sequestration strategies into soil stabilization processes and reduce greenhouse gas (GHG) emissions related to this process.
- **Climate Resilience:** The development of stabilization techniques that improve climate resilience is becoming more and more important as soil behavior and performance are impacted by climate change. This entails dealing with problems like temperature swings, elevated moisture content, and soil erosion. The goal of research is to develop stabilization strategies that support the long-term stability of infrastructure while being flexible enough to respond to shifting environmental conditions (Goetz, 2011; Marinelli et al., 2018).

10.1.4. Field Applications and Case Studies

- **Urban and Infrastructure Projects:** There is growing interest in the use of new soil stabilization methods in infrastructure and urban development projects. Case studies show how new techniques can be used to improve the performance of foundations, embankments, and roads. For instance, the application of biopolymers for slope stabilization and the use of recycled materials in road construction have demonstrated encouraging outcomes in improving soil qualities and sustainability.
- **Disaster Mitigation:** The use of soil stabilization techniques in efforts to mitigate disasters is growing. For example, in areas that are susceptible to landslides, slope stabilization and geosynthetics have been employed. Research is concentrated on assessing these methods' efficacy in lowering risk and boosting resilience as well as on optimizing them for quick deployment in disaster-prone areas.
- **Rural and Agricultural Applications:** New techniques for stabilizing soil are being investigated in agricultural and rural settings to enhance soil quality and promote sustainable land management strategies. The efficiency of methods like mechanical stabilization and the application of organic stabilizers in boosting soil fertility, halting erosion, and boosting agricultural output is being investigated.

10.1.5. Future Directions and Research Opportunities

- **Integration of Multidisciplinary Approaches:** Future investigations into soil stabilization are probably going to take a multidisciplinary approach, combining knowledge from environmental science, biology, engineering, and materials science. It will be crucial for researchers, practitioners, and industry stakeholders to work together to address complex challenges in soil stabilization and develop creative solutions.
- **Advanced Monitoring and Testing:** The performance of soil stabilization techniques must be assessed, and this requires the development of sophisticated testing and monitoring procedures. Real-time monitoring of soil behavior and treatment efficacy is becoming easier thanks to advancements in sensors, remote sensing, and data analytics. In order to increase accuracy and decision-making, future research will concentrate on incorporating these technologies into stabilization procedures.

- **Policy and Regulatory Considerations:** New policies and regulations will be required to guarantee the safe and efficient application of emerging soil stabilization techniques as they gain traction. Policy decisions, the development of new method guidelines, and possible standardization and quality control challenges will all be influenced by research.
- **Education and Training:** The education and training of experts in the field is essential for the effective implementation of newly developed soil stabilization techniques. Conducting research on optimal approaches for training and disseminating knowledge can facilitate the successful execution of novel techniques and guarantee that professionals are prepared to tackle changing obstacles in soil stabilization.

Recent developments in soil stabilization research are propelling important strides in the area and providing fresh approaches to improve resilience, sustainability, and soil performance. The future of soil stabilization is being shaped by ongoing research, which takes into account factors like field applications, environmental impact, and novel stabilizing agents and techniques. The construction industry can progress toward more sustainable and efficient soil management techniques by embracing these trends and tackling new issues. This will help to build resilient infrastructure and practice responsible stewardship of natural resources (Goetz, 2011; Marinelli et al., 2018).

10.2. CHALLENGES AND OPPORTUNITIES IN MODERN ROAD CONSTRUCTION

The road construction industry of today operates at the nexus of changing societal needs, environmental concerns, and rapid technological advancements. Road construction must navigate a complex landscape of challenges as cities grow and transportation demands rise while taking advantage of new opportunities to improve infrastructure sustainability, safety, and efficiency.

10.2.1. Challenges in Modern Road Construction

1. **Environmental Impact:** Reducing the environmental impact of road construction is one of the biggest challenges. Conventional methods of building roads frequently cause major ecological disturbances, such as pollution and habitat destruction. Carbon emissions and resource depletion are caused by the use of non-renewable resources like concrete and asphalt. Furthermore, construction activities can increase runoff, contaminate water, and erode soil, which increases

the risk of flooding and degrades the quality of the local water supply.

2. **Urbanization and Space Constraints:** There is less and less room for new road construction as metropolitan areas expand. In densely populated cities, expanding or improving road networks necessitates careful planning to prevent upsetting communities and existing infrastructure. This frequently entails intricate logistics, such as finding your way around utilities, historical sites, and existing buildings. The requirement for underground or vertical expansion adds expenses and new engineering challenges.
3. **Traffic Management and Congestion:** Road construction often makes traffic congestion worse, which affects local businesses and commuters. To reduce interruptions, traffic flow management during construction calls for creative solutions. Managing construction schedules to avoid peak traffic times and putting effective detour plans into action are difficult but necessary tasks. In addition, traffic congestion brought on by construction projects may worsen air quality and cause anxiety in the drivers involved.
4. **Safety and Worker Welfare:** Because road construction is such a hazardous job, safety is still a major concern. Accidents involving large machinery, falling objects, and loud noises are common at construction sites. Wearing protective gear, conducting frequent training sessions, and strictly adhering to safety procedures are all necessary to ensure worker safety. Additionally, because construction work is demanding and can result in stress and burnout, it is important to address the mental health and welfare of workers.
5. **Budget and Resource Management:** Overspending on the budget is a typical problem in road construction projects. Financial resources may be strained by unforeseen expenses brought on by changes in material prices, unforeseen site conditions, and project delays. To keep expenses under control and guarantee that projects are finished on schedule and within budget, effective project management and planning are essential. The effective management of resources, including the availability of qualified personnel and supplies, is essential to the accomplishment of projects.

10.2.2. Opportunities in Modern Road Construction

1. **Technological Advancements:** There are many opportunities to enhance the road construction process thanks to technological innovations. Cutting-edge materials that improve durability and

lessen environmental impact include sustainable concrete and recycled asphalt. More accurate planning and execution are made possible by the use of smart construction technologies, such as drones, 3D modeling, and real-time data analytics. Robotics and automated equipment can expedite construction work, boosting productivity and security.

2. **Sustainable Practices:** There are chances to lessen environmental footprints as road construction becomes more sustainable. More environmentally friendly projects are the result of green construction techniques, such as the use of low-impact materials and energy-efficient building techniques. Permeable pavements and urban green spaces are two examples of green infrastructure solutions that can be used to lessen the negative environmental effects of road construction. Sustainable development objectives are further supported by the incorporation of renewable energy sources, such as solar-powered streetlights.
3. **Smart Infrastructure:** For contemporary road building, the idea of smart infrastructure offers a huge opportunity. Real-time monitoring and management are made possible by the integration of sensors, data gathering systems, and communication technologies into road networks. With automated systems, smart roads can improve safety, identify road conditions, and deliver real-time traffic information. Because of this connectivity, traffic management is made easier, congestion is lessened, and overall road safety is increased.
4. **Enhanced Materials and Techniques:** The advancement of cutting-edge building materials and methods presents chances to enhance the lifespan and performance of roads. Roadway lifespans can be increased by innovations like high-performance concrete, which provides superior durability, and self-healing asphalt, which fixes small cracks on its own. Furthermore, methods like modular road sections and 3D printing of building components can lower costs and expedite the construction process.
5. **Community Engagement and Social Benefits:** The advantages of road construction projects can be increased by interacting with communities and taking their needs into consideration. Participating in the planning process with local stakeholders guarantees that road designs take into account local concerns and goals. Green spaces and bike lanes are examples of pedestrian-friendly features that can enhance the quality of life for locals in projects. Additionally, by boosting local businesses and creating jobs, road construction projects frequently contribute to the economy.

6. **Resilience and Adaptability:** Creating roads that are resilient to severe weather and shifting climatic conditions is a crucial chance to future-proof infrastructure. Creating resilient road systems with adaptable elements like better drainage and materials resistant to flooding aids in reducing the effects of climate change. Furthermore, it is easier to adapt roads to changing transportation needs and future technological advancements when they are designed with flexibility in mind.

The contemporary road construction sector faces a number of difficulties, including budget management, safety, and space and environmental impacts. These difficulties do, nevertheless, also offer chances for creativity and advancement. Road construction techniques can be advanced while addressing these issues through the use of enhanced materials, smart infrastructure, sustainable practices, and technological advancements. By seizing these chances and emphasizing community involvement and resilience, the road construction sector can help create future transportation networks that are safer, more sustainable, and more effective (Goetz, 2011; Marinelli et al., 2018).

10.2.3. The Challenges of Road Management Towards Sustainability and Development

There are numerous significant obstacles in the way of sustainable and development-oriented road management. Since road construction and maintenance frequently result in habitat destruction, increased pollution, and resource depletion, environmental impact is a major concern. Degradation of the environment and carbon emissions are caused by traditional materials and methods. Incorporating eco-friendly materials and sustainable construction techniques—which can be more expensive and unfamiliar to industry stakeholders—is necessary to address these problems.

Road management is made more difficult by urbanization and space constraints. It becomes difficult to upgrade or expand road networks in densely populated areas as cities grow. The requirement to maneuver around historical sites and existing infrastructure necessitates careful planning and creative solutions, which frequently raises project complexity and costs.

Financial constraints are yet another important issue. Because sustainable road management uses cutting-edge materials and technologies, initial costs are frequently higher. Finding practical, long-term solutions at a reasonable cost and carefully managing finances are required to balance these expenses with financial limitations. Roads must be maintained and adapted to in order to continue being useful and sustainable over time. While addressing wear and tear requires routine maintenance, implementing sustainable practices can be more costly and logistically challenging. Furthermore, continued investment and innovation are needed to adapt roads to changing climatic conditions and future technological advancements. Public Awareness and Engagement are also essential. In order to manage roads sustainably, the community must be understood and supported. Gaining public support and facilitating the easier implementation of new initiatives can be accomplished by educating the public about the advantages of sustainable practices and involving them in decision-making processes.

Road networks continued to expand globally in the early decades of the 21st century, in spite of funding shortages and harsh criticism from environmentalists. The development of the road network and associated transportation activities promotes economic growth, social welfare, and job opportunities. By 2050, this trend is predicted to create about 25 million km of new roads, the majority of which will be in developing nations (Campbell et al., 2017). It is well known that having a functional road network can significantly impact both inclusive economic growth and the reduction of poverty. The Sustainable Development Goals for 2030, which establish goals for enhancing access to transportation and advancing the caliber, dependability, security, sustainability, and resilience of roads, bridges, and other infrastructure, represent this role.

Investments in road transportation have direct advantages in terms of reduced travel times and expenses as well as the creation of temporary jobs, mostly in the construction industry. It appears that these investments indirectly serve as facilitators since improved accessibility boosts productivity, which in turn promotes trade and fosters regional cohesion by lowering transportation costs (Goetz, 2011). It must be made very clear, though, that while road transportation is a signpost for regional development and economic growth, it is not a reliable indicator of an expanding or thriving economy in and of itself. The health of the economy at the regional or national level is determined by the rate at which public and private investments are made in industry, commerce, tourism, services, and agriculture, as well as by the sustainable use of local resources.

The road transport industry plays a significant role in the economy, both directly (through the transportation of goods, industrial products, and raw materials) and indirectly (through the creation of social and national wealth, for example) as well as through supporting other economic activity. Among all modes of transportation, road transport contributes the greatest value, estimated to be between 3 and 5% of the global Gross Domestic Product (GDP).

10.2.4. Road Transport and Technological Progress

The road industry plays a vital role in the modern world economy and lifestyle by meeting the growing demand for freight and mobility worldwide. One of the most important requirements for a robust economy is the smooth running of the roads and transportation sector, which enhances accessibility and connectedness and offers middle-tier services to ease communication between consumer centers and production facilities.

Technological advancement and innovation have been the main drivers of the boom in the road construction industry. New products, tools, materials, and methods have been brought to the road engineering sector. The building of earthworks, viaducts, tunnels, and bridges has been made easier by the use of cutting-edge methods and strong, contemporary machinery. New road connections, particularly freeways, ring roads, and international corridors, are now completely operational and provide exceptional serviceability and safety. These indisputable facts could support the logical conclusion that there aren't any issues or defects in the field of road engineering at the moment. This conclusion, however, hardly accurately captures the state of affairs in the majority of developing and developed nations. Numerous economies were severely impacted by the economic crisis, which made it difficult to finance both new projects and the modernization of existing connections. Significant direct, indirect, and cumulative effects on the environment have been linked to the documented growth of road transport operations—issues that have grown in importance in recent years. The density of transportation infrastructure in many Western European nations leaves insufficient room for other forms of production.

Furthermore, despite considerable progress in many European countries, road safety is always at risk due to climate threats, which continue to pose serious issues for both developed and developing nations. In this sense, significant problems pertaining to network operations and infrastructure integrity are a constant source of concern for policy stakeholders, local and federal governments, and road operators. These are genuine obstacles in a world that is constantly changing (Goetz, 2011; Marinelli et al., 2018).

10.2.5. Road Infrastructure Challenges

Numerous significant obstacles affect road infrastructure's sustainability, safety, and efficacy. An important problem with aging infrastructure is that many roads and bridges that were constructed decades ago now need major maintenance or replacement. Structural failures are caused by weather, heavy use, and poor maintenance, which deteriorates the structure and increases repair costs and safety hazards. The second major issue is traffic congestion. Road networks frequently become overloaded as urban areas grow and the number of vehicles increases. This causes delays, longer travel times, and higher pollution levels. To reduce traffic, efficient infrastructure planning and traffic management are essential, but putting these strategies into practice can be expensive and difficult.

Road infrastructure management is made even more difficult by funding and budgetary constraints. Numerous projects encounter financial constraints, as inadequate funds are allotted for essential upgrades and upkeep. The quality of infrastructure as a whole is often impacted by delays and project scalebacks brought on by budget overruns and the high cost of contemporary materials and technologies. Environmental Impact is being examined more closely. Conventional methods of building roads exacerbate pollution, resource depletion, and habitat destruction. Sustainable materials and practices are becoming more and more necessary, but incorporating them into current systems can be difficult and costly.

The integration of technology brings with it both advantages and disadvantages. Even though cutting-edge technologies can improve traffic efficiency and safety, integrating them into the current infrastructure will cost a lot of money and expertise. Managing the transition and guaranteeing compatibility with older systems can be challenging. Road accidents and maintenance-related risks put workers and the general public at risk, so safety concerns are of the utmost importance. To address these and guarantee road user safety, strict safety procedures and creative solutions are needed. In order to create a road network that is more efficient and sustainable, these issues must be balanced and address issues such as aging infrastructure, traffic congestion, funding limitations, environmental effects, technological integration, and safety concerns.

10.2.5.1. Financing Road Construction Projects

Both local and national economic development depend on roads. From this angle, the construction of major highways and road links is intended to promote development and expansion. But roads are also necessary for social cohesion, providing access to health, education, and service facilities in areas and provinces where profit is not clearly anticipated. It can be difficult to finance road construction projects with very little return, particularly in developing nations. While it is not as difficult, financing construction projects is still a challenge when it comes to toll motorways. Building toll roads has historically been a profitable venture, but during recessions, funding may be scarce or nonexistent. Stronger economies may be able to effectively address the issue, but they will have a harder time obtaining the funding needed for road construction projects, which are the engines of economic growth, in weaker nations.

10.2.5.2. Energy Consumption in Transport

Transportation continues to be reliant on oil; approximately 94% of total energy consumption comes from fuels derived from oil. Three-quarters of the total energy used in transportation is accounted for by road transport, which uses the most energy. Compared to other forms of transportation, the energy consumption of road operations continues to increase. The percentage of diesel fuel used in road transportation in Europe increased steadily from 52% in 2000 to over 70% in 2014. This increase is indicative of the continent's vehicle fleet becoming more and more dieselized over time.

Furthermore, highway infrastructure requires energy on its own. In particular, the highway network needs energy to run numerous data collection and digital information hardware devices, as well as for lighting and signage. Because road operations use a lot of fossil fuels, they pose a serious energy challenge to road transportation. This problem is not sustainable and calls for immediate alternative solutions.

10.2.5.3. Road Serviceability

Road serviceability is a broad term that encompasses service stations, traditional and electronic information, pavement ride quality, and safety level. A broader definition of serviceability may encompass both road capacity and service level. Travelers and drivers expect the highest caliber of road performance and anticipate enhanced serviceability in the future. Without a doubt, this calls for both human and financial resources, not only for the on-site operations but also in order to plan efficient engineering interventions and anticipate maintenance needs.

The European Commission's Eurobarometer survey on the quality of transportation in the EU found that 60% of respondents ranked traffic congestion and 59% ranked road maintenance as the two most important issues facing Europe's roadways. In fact, traffic congestion and road defects are two external costs of road infrastructure that are primarily borne by drivers; these factors have an impact on the road network's dependability and service quality.

The complicated phenomena of congestion are impacted by a variety of technical, socioeconomic, and even human factors. It can be addressed with a variety of strategies, such as expanding the road, building detours, and switching to public transportation. In many European road networks, road deterioration and pavement distress necessitate maintenance planning and financial outlays.

Road serviceability is more important than national and regional networks in industrialized and developing nations where motorway networks are being developed, as funding maintenance or rehabilitation projects on these networks may be a secondary concern. On the other hand, a toll highway in need of clearly urgent maintenance may be degraded due to a lack of resources (Figure 10.1).

10.2.5.4. Climate Change Impacts on Roads

The road industry is susceptible to the effects of climate change. Extreme weather and climate change present serious obstacles to the sustainability, efficacy, safety, and dependability of road transportation networks. Natural disasters like hurricanes, wildfires, floods, and tsunamis pose a serious threat to the safety of people, property, and vehicles as well as the integrity of the transportation network. Since safe and dependable road transportation is a critical component of global economic development and social well-being, national transportation authorities and motorway operators need to make necessary infrastructure adjustments to accommodate climate change and improve road transportation's resistance to extreme weather.

Since the early 2000s, the majority of industrialized countries have shown genuine concerns about how to adapt their road infrastructure to extreme climate events. But even with the scientific development of risk assessment techniques and appropriate engineering solutions, it appears that climate threats have not been effectively addressed or mitigated up to this point. There is always a need for additional funding as well as for different, more efficient approaches.

10.2.5.5. Environmental Impacts

The escalating need for road transportation, travel options, and inventive transportation infrastructure has been linked to various environmental consequences, including exposure to noise and air pollution, altered land use, habitat fragmentation, and impacts on wildlife and biodiversity.

Furthermore, the scientific community tracking climate change is becoming increasingly concerned about lowering GHG emissions from transportation. Approximately 25% of the GHG emissions produced by the EU are attributed to transportation in Europe, where emissions have increased recently. The only significant economic sector in Europe where GHG emissions have increased since 1990 is the transportation sector. In particular, the transportation industry is a significant and rapidly expanding source of GHG emissions; in 2013, it accounted for nearly 73% of all transport-related GHG emissions. In this regard, several Western European nations have adopted alternative strategies to cut GHG emissions at the regional and local levels (Goetz, 2011; Marinelli et al., 2018).

Figure 10.1. *Poor road serviceability due to surface cracks: Urgent need for maintenance on toll motorway.*

Source: By Bidgee – Own work, CC BY-SA 3.0, https://commons.wikimedia.org/w/index.php?curid=7367880.

10.2.5.6. Safety on Roads and Highways

The World Health Organization estimates that traffic accidents result in up to 50 million injuries and 1.25 million deaths annually, or roughly 2,400 road fatalities. Among people aged 15 to 29, traffic accidents are the main cause of death. In addition to proposing measures to address the global road safety crisis, the UN has passed many resolutions on the subject of road safety. Globally, the Decade of Action for Road Safety 2011–2020 has sparked a lot of activity. One of the Sustainable Development Goals is to reduce traffic-related deaths and injuries by 50% by the year 2020. In OECD/high-income countries, there has been significant progress in the past 10 years in lowering the death toll; however, the situation in low-income countries is different. Furthermore, it appears that there hasn't been much attention paid to the problem of pile-up accidents on motorways. Forecasts for traffic fatalities indicate that there will be a considerable increase in road deaths worldwide, thereby highlighting the ongoing challenge of road safety.

10.2.5.7. Economy and Cost-Effectiveness

In short- and medium-distance travel operations, road transport holds a prominent position among all modes of transportation. When compared to other forms of transportation, the cost of transportation (per ton × km) is still high and only becomes more expensive as the distance traveled increases. Road transport costs are made up of two categories: external costs (noise, congestion, infrastructure damages, health and environmental issues) and direct costs (fuel, capital depreciation, maintenance, motorway tolls, ferry fares, and wages).

Similarly, driving is typically an expensive option, regardless of how appealing it may be for other reasons. Road closures, delays, and congestion can result in a large increase in direct and indirect costs. Road operators work to lower travel costs for the benefit of motorists, transit organizations, and the environment by putting cutting-edge technology into practice.

10.2.6. Responses and Remedies

10.2.6.1. Unsolicited Project Proposals and Value Engineering

When public resources are scarce, public authorities typically request proposals for projects that would be best carried out through the Public-Private Partnership (PPP) modality. Solicited proposals are those that are submitted by private parties in response to this kind of request. Private parties may occasionally submit proposals without the public authorities requesting them. We refer to these offers as unsolicited project proposals. Many nations have implemented unsolicited PPP projects; however, some do not accept such proposals due to ethical concerns, particularly the risks associated with competition and transparency (United Nations Economic and Social Commission for Asia and the Pacific, or ESCAP).

These kinds of proposals frequently stem from creative project concepts. The challenge with unsolicited proposals, however, is striking the correct mix between promoting creative project ideas from private enterprises and maintaining the openness and efficiency benefits of a competitive tender process. Some governments have created systems to turn unsolicited proposals into competitively tendered projects, giving the proponent an advantage in the bidding process, in recognition of the merits of unsolicited proposals that they frequently receive. Value engineering also seems to be a novel idea that the authorities in charge of tendering should consider. The idea is to maximize an asset's or project's performance-to-cost ratio (Marinelli et al., 2018) by either cutting costs or raising performance. Though value engineering is much more than this, it can be thought of as an extension of a cost-benefit analysis. It is the process of finding non-obvious, alternative, and cost-effective solutions while also fully utilizing the asset to maximize its value.

Because of the heavy traffic volume, the problem of expanding road capacity frequently comes up in the context of urban or peri-urban road infrastructure. The Ring Road in Thessaloniki, Greece, needs to be significantly widened due to its congestion. Widening the Ring Road would be nearly impossible due to its location between an urbanized area and a forestry zone. The first, and most expensive, option was to construct a 2 × 2 motorway on a new alignment through a tunnel. A cross-cantilever bridge (an example is in Figure 10.2) appears to be equally functional by using a "think different" approach; it would only cost 25% of the price of the tunnel solution and take less time to build (Goetz, 2011; Marinelli et al., 2018).

Figure 10.2. *An example of a cross-cantilever bridge.*

Source: By Mr Master Keyboard – Own work, CC BY-SA 4.0, https://commons.wikimedia.org/w/index.php?curid=121262762.

10.2.6.2. The Energy-Harvesting Road

Energy-efficient transportation needs to be encouraged on three different levels: system efficiency, travel efficiency and vehicle efficiency.

Corresponding to these three levels of energy efficiency in transport, three basic strategies exist to improve energy efficiency:

- Avoiding increased transport activities and reducing the current demand for transport
- Shifting demand for more efficient modes of transport
- Improving the vehicles and fuels used

Furthermore, highways and the infrastructure that goes along with them may present chances for energy production. The land that is directly owned by highways and associated with them is a significant asset in terms of potential energy generation. The 2014 PIARC Report outlines the various energy-generating technologies that are available for use in the context of highways and evaluates the potential to lower the energy consumption of the current highway infrastructure. These technologies allow highway facilities to produce energy or use less energy. They include photovoltaic collection, thermal collection using specific pavement appliances, wind micro-generation, and LED lighting solutions. All of these technologies are still in the experimental stage, though.

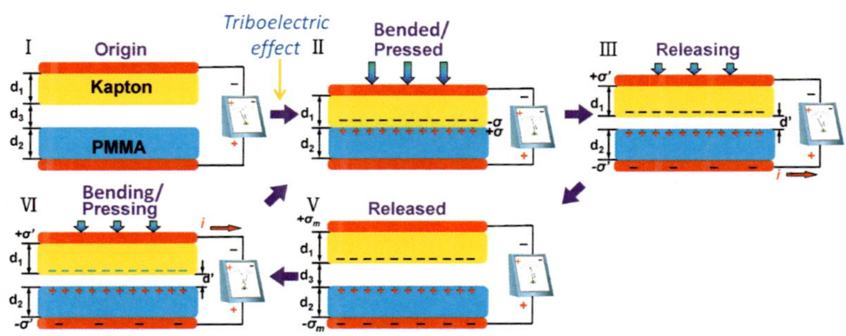

Figure 10.3. *Laboratory device for triboelectric energy generation.*

Source: By Wangsh05 – Own work, CC BY-SA 3.0, https://commons.wikimedia.org/w/index.php?curid=29499503.

Triboelectric energy generation is still being studied at Aristotle University of Thessaloniki (AUTH). The effect of a rolling wheel on an asphalt layer has been shown to produce low-intensity electric current in laboratory tests on asphalt surfaces and electrical conductor settings (an example is in Figure 10.3)

(Barkas et al., 2019). This study aimed to specify and create an assembly that will be placed on a pavement surface within an urban road network, to generate electricity and recharge nearby street lights.

10.2.6.3. Recycled and Re-Utilized Pavement Materials

The process of incorporating recycled materials into pavement construction involves several different aspects. Undoubtedly, this practice has a negative impact on the environment, but it's also critical to understand the engineering and financial implications of recycling in pavement construction. The use of recycled materials in the highway industry as a whole needs to be increased, according to the Federal Highway Administration (FHWA).

There are several reasons to develop a recycling-friendly strategy in road construction:

- Cost-savings potential
- Life-cycle cost and engineering performance
- Reduction in landfill
- Stewardship of our environment

Currently, a wide range of materials—including fly ash, steel slag, glass, rubber tires, demolition debris, old asphalt layers, and others—are recycled when building new pavement. The majority of them perform perfectly well as replacement pavement materials; however, the primary disadvantages for a variety of applications appear to be the expense of transportation and the requirement for pre-processing.

Research on the potential repurposing of industrial by-products, particularly fly ash, steel slag, and bauxite residue, is currently underway at Aristotle University. The viability and potential of using recycled plastic in pavements have also been studied more recently. The integration of these recyclable materials in road structures has always been tested in pilot and actual construction projects, and laboratory research has produced encouraging results. A hydraulic binder called fly ash can be utilized to build semi-rigid pavements, and steel slag aggregates are a component of antiskid asphalt layers. The practice of repurposing fly ash and steel slag in road structures is actually common in some European countries, primarily due to resource scarcity in certain areas. Aristotle University's research and pilot projects on the use of bauxite residue as a building material demonstrate the extraordinary qualities of this byproduct and produce highly durable, strong, and high-quality road engineering structures.

10.2.7. Strategies for Climate-Change Adaptation of Road Infrastructure

Over the past few decades, notable advancements have been made in the adaptation of road infrastructure to extreme weather events. The national road authorities in Europe and other developed nations have faced a constant challenge in adapting their roads to the changing climate since the early 2000s. The concern over climate hazards has gradually grown, and national and international action plans have been developed in response. It has been reported that the implementation of protection measures, particularly against flooding, in several countries has resulted in positive effects on network integrity.

It must be emphasized, nevertheless, that the industrialized world's current state of affairs is far from ideal or even appreciably better. In Europe and beyond, severe wildfires, catastrophic floods, and copious amounts of snowfall frequently occur without prompt and efficient responses. One can easily feel helpless when viewing photos of catastrophic weather events. Finding engineering adaptation strategies during a network that has been severely damaged can occasionally be seen as a fruitless endeavor.

Figure 10.4. *Wildlife overpasses in Singapore.*

Source: https://en.m.wikipedia.org/wiki/File:Wildlife_overpass_in_Singapore.jpeg.

However, research efforts and a strong commitment in this direction must continue despite the occasional inadequacy or ineffectiveness of engineering measures to combat climate threats. At Aristotle University, a somewhat different strategy has been developed to counteract extreme weather events and fortify road infrastructure. This method proposes alternative engineering courses of action, like bringing the road link's initial balance back to a larger contextual scale, developing workable solutions and value-engineering strategies, ensuring emergency services are ready and prepared, and utilizing new technologies to forecast and anticipate impending risks. When it comes to climate threats, all of them could be considered formidable and potent weapons.

10.2.7.1. A Low-Carbon and More Sustainable Road Infrastructure: The Green Highway

It is essential to comprehend the basic connections between road activities, infrastructure, and the physical environment (topology, geology, climate, location, hydrology, natural vegetation, and animal life). From an economic, social, and environmental standpoint, sustainable road transport planning considers the Green Highway as a potential solution.

The model's scope is to diminish energy needs, road accidents and interventions, hence restoring operability and serviceability using low-cost techniques.

The model is articulated over three different sets of measures:

- Energy consumption, including auto-fluorescent markings, auto-dimming lighting and solar islands at rest areas.
- Road safety and serviceability, including rumble strips on straight segments, asphalt anti-icing additives and perpetual asphalt pavements.
- Geotechnics and environment, including borrowed-soil spraying on rock slopes, versatile wood screens and balancing ponds.

10.2.7.2. Real-Time Safety Management System

The most creative uses of digital technology are warning systems and real-time safety condition assessments. These are systems that take in and process real-time images of traffic conditions, analyzing them to determine the likelihood of an accident. They appropriately alert and warn drivers about the current state of road safety through variable message signs or in-car displays (connected vehicles). On motorways, there are many factors that can cause an accident, including off-lane border driving, driving carelessly, abandoned cars, and objects on the road. When the system detects potentially dangerous circumstances, the digital software sends out a warning signal. The implementation of these systems could be advantageous when it comes to clogged motorways and black spots on public roads.

10.2.7.3. Transition to Infrastructure and Logistics for Electric Vehicles

The development of electric vehicles is a subject that is progressively receiving more attention in the field of infrastructure for road transportation. From an initial perspective, the cutting-edge technology of electric cars appears to solve the pressing problem of pollution in cities. But the technology's economic effects—whether they be individual or collective—will also be significant. It is commonly known that compared to fuel costs for conventional vehicles, the electricity costs per mile for plug-in electric vehicles, or PEVs, in charge-depleting mode, are likely to remain significantly lower and less volatile. Furthermore, even though an electric vehicle's reduced environmental impact is the primary motivator to buy one, it appears that financial considerations will weigh more heavily in the decision than environmental protection-related ones.

Road infrastructure that is conducive to the use of electricity will primarily depend on how far the automotive industry advances in terms of cost and vehicle autonomy. The technical limitations associated with batteries and plug-in settings will undoubtedly be gradually resolved by industrial research, but the real challenge lies in providing electric power produced by renewable energy sources and gradually reducing or eliminating the use of fossil fuels. The two main things that drivers look for in a road infrastructure are safety and comfort. Road infrastructure is always changing to incorporate new technological advancements for the benefit of drivers, passengers, and automobiles. The focus of these recent technological advances is on intelligent transportation systems, high-quality materials, and safety equipment. Road authorities must now address three new issues: the decarbonization of road transportation, environmental restrictions, and electronic information, despite the significant steps made toward a technological and engineering boost in road construction. Global transportation problems are generally significant and critical, and failure to take action or propose solutions could lead to a growing economic divide between developed nations and developing ones (Goetz, 2011; Marinelli et al., 2018).

10.3. FUTURE DIRECTIONS FOR INNOVATION IN SOIL STABILIZATION TECHNIQUES

An essential component of civil engineering is stabilizing the soil, especially when building roads, foundations, and other infrastructure. Enhancing the engineering qualities of soil, such as its strength, durability, and workability, is the main objective of soil stabilization to support the structures that are built upon it. Although effective, traditional techniques such as mechanical compaction, chemical stabilization with cement or lime, and the use of geosynthetics are coming under increasing pressure to meet new standards for resilience, efficiency, and sustainability. To address these issues and progress in the field, several cutting-edge approaches to soil stabilization are beginning to emerge.

10.3.1. Advanced Materials and Technologies

The creation of novel materials and technologies is one of the most fascinating fields of innovation. Researchers are looking into using novel polymers and bio-based materials like biochar to improve the characteristics of soil. Because it can hold more water and prevent soil erosion, biochar—a type of charcoal made from organic material—has shown promise in enhancing soil stability. In a similar vein, the potential of polymers—including biopolymers made from natural sources—to bind soil particles better than conventional techniques is being researched. In addition to increasing soil stability, these cutting-edge materials may have positive effects on the environment by lowering carbon footprints and dependency on non-renewable resources.

10.3.2. Nanotechnology

10.3.2.1. Soil Stabilization Through a Cutting-Edge Approach: The Use of Nanotechnology

The physical and chemical characteristics of soil can be drastically changed by applying nanoparticles. As an example, nanoparticles can be used to strengthen soil, increase resistance to water infiltration, and improve soil compaction. Materials that can be used to alter soil properties at the molecular level and produce more successful stabilization include nano-silica and nano-clay. The full potential and implications of nanotechnology in soil stabilization, including its effects on the environment and long-term stability, are still being investigated.

10.3.3. Geosynthetics and Hybrid Systems

Although soil stabilization has long employed geosynthetics, developments in this area continue to push the envelope. Improved geosynthetic materials are being developed, including geomembranes, geotextiles, and geogrids. To achieve better performance, these materials can be combined to create hybrid systems that incorporate various geosynthetic types. For example, geotextiles and geogrids can be combined for improved soil reinforcement and load distribution. Another example of geosynthetic innovation is the creation of intelligent materials that adapt to environmental changes, such as fluctuations in temperature or moisture content, to maintain ideal stabilization conditions.

10.3.4. Biotechnological Approaches

The use of biotechnological techniques is becoming more and more viable for stabilizing soil. By using bacteria to encourage the formation of calcium carbonate crystals in the soil, a process known as "microbial-induced carbonate precipitation" (MICP) effectively binds soil particles together. This procedure can greatly decrease permeability and increase soil strength. To improve soil stability, scientists are also looking into using biological agents such as plant roots. For example, certain fungi can enhance soil structure and nutrient availability, and deep-rooted plants can stabilize slopes and stop erosion.

10.3.5. Sustainable and Environmentally Friendly Methods

In soil stabilization techniques, sustainability is becoming more and more important. The environmental consequences of traditional methods, like the use of cement and lime, include high carbon emissions and resource depletion. New strategies concentrate on using waste materials and by-products to lessen the impact on the environment. For instance, using construction debris and recycled asphalt for soil stabilization not only lowers waste but also offers a practical solution at a reasonable price. Furthermore, integrating low-impact and renewable resources into soil stabilization techniques is consistent with the more general objectives of environmental sustainability and green construction methods.

10.3.6. Advanced Testing and Monitoring

The practices of soil stabilization are also changing as a result of the advancement of sophisticated testing and monitoring technologies. More advanced techniques like in-situ sensors and remote sensing technologies are being used to supplement traditional testing methods like soil compaction tests and shear strength tests. These developments make it possible to monitor soil conditions in real-time, giving important information for modifying stabilization techniques and guaranteeing peak performance. The use of geophysical methods and drone-based surveys to swiftly and precisely evaluate soil properties over wide areas is one of this field's innovations.

10.3.7. Computational Modeling and Simulation

The future of soil stabilization greatly depends on computational modeling and simulation. The creation of intricate models that can forecast soil behavior under varied circumstances is made possible by improvements in software and processing power. By simulating the effects of various stabilization techniques, these models can assist researchers and engineers in refining their strategies before implementation. In order to improve decision-making and prediction accuracy, these models are also incorporating artificial intelligence and machine learning. These technologies can be used to create more effective and efficient soil stabilization solutions that are suited to the requirements of particular projects.

10.3.8. Integration with Sustainable Infrastructure Practices

Future innovation should focus on integrating soil stabilization with more comprehensive sustainable infrastructure practices. One way to achieve this is to match stabilization methods to green building guidelines, like the ones provided by the Leadership in Energy and Environmental Design (LEED) certification. For example, techniques for stabilizing soil that improve stormwater management and lessen erosion while also adding to the overall sustainability of a construction project are becoming more and more valued. Furthermore, integrating soil stabilization into more comprehensive environmental management plans, like habitat restoration and land reclamation, can increase the overall influence and advantages of stabilization initiatives.

10.3.9. Resilience to Climate Change

Soil stabilization techniques need to adjust to the increasing variability of environmental conditions brought about by climate change. This entails dealing with problems like intensified rainfall, temperature increases, and an increase in the frequency of extreme weather events. Future developments in soil stabilization must concentrate on making infrastructure and soil more resilient to these shifting circumstances. For instance, managing the effects of heavy rainfall and flooding will require strategies that increase soil drainage and decrease erosion. In a similar vein, techniques to improve the thermal characteristics of soil can aid in reducing the impact of temperature variations on soil stability.

10.3.10. Collaboration and Interdisciplinary Research

There's a good chance that more interdisciplinary research and collaboration will influence soil stabilization in the future. The field of soil stabilization is complex and involves relationships with environmental engineering, biology, materials science, and other areas. To drive innovation and put new techniques into practice, researchers, practitioners, and industry stakeholders must work together. Multidisciplinary study can result in the creation of comprehensive solutions that handle various facets of soil stabilization and support more efficient and long-lasting methods. There are many interesting and promising avenues for soil stabilization technique innovation in the future.

The development of soil stabilization techniques is being aided by developments in materials science, nanotechnology, geosynthetics, biotechnology, sustainability practices, testing and monitoring, computational modeling, and climate change resilience. The performance, sustainability, and environmental impact of soil stabilization techniques could all be enhanced by these developments, which could ultimately result in more resilient and efficient infrastructure. To achieve these developments and take advantage of the opportunities and challenges that the field of soil stabilization will present, cooperation and multidisciplinary research will be essential (Goetz, 2011; Marinelli et al., 2018).

REFERENCES

1. Barkas, D. A., Psomopoulos, C. S., Papageorgas, P., Kalkanis, K., Piromalis, D., & Mouratidis, A. (2019). Sustainable energy harvesting through triboelectric nano-generators: A review of current status and applications. *Energy Procedia, 157*, 999–1010.
2. Boehler-Baedeker, S., & Hueging, H. (2012). *GIZ Sourcebook on Urban Transport and Energy Efficiency*. German Cooperation Agency (GIZ).
3. Campbell, M., Alamgir, M., & Laurance, W. (2017). Optimising future tropical roads: Examining the economic benefits and environmental costs of roads in the Asia-Pacific. *Australian Wildlife, 1*, 26–29.
4. European Commission. (2014). *Quality of Transport*. EC Report, Special Eurobarometer 422a. Brussels, Belgium: European Commission.
5. Goetz, A. (2011). The global economic crisis, investment in transport infrastructure and economic development. In K. Button & A. Reggiani (Eds.), *Transportation and Economic Development Challenges* (pp. 42–71). Edward Elgar.
6. Kehagia, F. (2008). Skid resistance performance of asphalt wearing courses with electric arc furnace slag aggregates. *Waste Management Research, 27*, 288–294.
7. Marinelli, M., Petroutsatou, K., Fragkakis, N., & Lambropoulos, S. (2018). Rethinking new public infrastructure value for money in recession times: The Greek case. *International Journal of Construction Management, 18*(4), 331–342.
8. Mouratidis, A., & Pernientaki, I. (2018). Engineering assessment of a local road constructed from bauxite residue. *9th International Conference on Waste Management and the Environment*, 101–109.

INDEX

D

E

F

G

Q

R

S

T

U

FSC
www.fsc.org
MIX
Papier | Fördert gute Waldnutzung
FSC® C083411

Kolibri 360 GmbH
Ferdinand-Jühlke-Straße 7
99095 Erfurt, Deutschland
produktsicherheit@kolibri360.de